DYNAMICAL SYSTEMS IN THE PLANE

DYNAMICAL SYSTEMS IN THE PLANE

Otomar Hájek

ACADEMIC PRESS / London and New York
1968

ACADEMIC PRESS INC. (LONDON) LTD
BERKELEY SQUARE HOUSE
BERKELEY SQUARE
LONDON, W. 1.

U. S. Edition published by

ACADEMIC PRESS INC.
111 FIFTH AVENUE
NEW YORK, NEW YORK 10003

Library of Congress Catalog Card Number: 67-28007

Printed in Czechoslovakia by Academia, the Publishing House of the Czechoslovak Academy of Sciences

Preface

The present book is the result of an attempt to develop into a coherent theory, and place into appropriate context, several results on dynamical systems which have appeared earlier in separate papers. To this purpose, the editors of the following journals have given their kind permission: Commentationes Mathematicae Universitatis Carolinae (as concerns Hájek 4, 9 and 10, 1964, 1965 and 1965); Collections – Technical University of Prague (Hájek 8, 1965); Czechoslovak Mathematical Journal (Hájek 5 and 6, 1964 and 1965).

The author wishes to express his indebtedness to several of his colleagues for the benefit of discussions and criticism, and in particular to Mr Josef Nagy who read in detail several versions of the text. Mr David Carlson and Mr Roger McCann have helped in the terminal stages of the work. As far as the author is concerned, there is little doubt that the book would not have been written without the original impetus, and support during the preparation of the manuscript, from Academic Press.

Caroline University, Prague — OTOMAR HÁJEK

Contents

Introduction

There are three aspects of dynamical system theory which it seems appropriate to separate out; the reader may, according to this inclination, prefer to emphasise one or the other of these.

Historically, the subject is an outgrowth of the qualitative theory of differential equations, and this latter has applications, or at least reasonable interpretations, in the study of the behaviour in-the-large of an extensive class of physical systems. More specifically, there are the familiar connections with theoretical dynamics and non-linear mechanics; and in mentioning some of the more distinguished names, those of Poincaré, G. D. Birkhoff, and van der Pol, Minorski can hardly be omitted.

Secondly, dynamical system theory may well be considered as applied topology, and indeed general, analytic and algebraic topology have all played a fundamental role throughout the development of the subject. The author's preference of this view-point will no doubt be apparent, through not intentionally so. In any case, this aspect will be the excuse for making free use of previously established topological results. Thus, no attempt is made to set up the exposition in a manner independent, formally at least, of other treatises.

Finally, it is proposed that dynamical system theory is an independent mathematical discipline, with its own subject-matter, proof methods and fundamental results. It is hoped that the present treatise will suggest that dynamical systems can no longer be considered as a mere chapter in the theory of differential equations; and that the results established may form a basis for further developments of the theory.

To borrow an introductory sentence from Gottschalk and Hedlund (1955), "It was Poincaré who first formulated and solved problems of dynamics as problems in topology". Once this approach had become familiar, another moment manifested itself, that one was studying the (topologically) invariant properties of a non-invariant object, viz. a system of differential equations. A. A. Markov performed the logical next step of defining an essentially topological concept, containing as special case the (autonomous) differential equations. This was called a dynamical system or, in a different terminology, a transformation group of a metric space with the reals as coefficient group.

It is then natural to replace the metric space by a general topological space as carrier of the dynamical system; and in order to obtain an even closer correspondence with possibly non-prolongable differential equations, a modification of this concept was introduced in Hájek 8 (1965) and termed a *local dynamical system*. The purpose of the present treatise is then the study of continuous local dynamical systems on topological spaces, with special reference to those properties which are best applicable to 2-manifolds as carriers. In Chapter II there are isolated several results which are purely dynamical in the sense of being independent of any continuity structure on the carrier. Chapter IV then treats the elementary consequences of the co-existence of mutually compatible dynamical and topological structures on the same set; these two chapters constitute the fundamentals of dynamical system theory. As in similar cases, it is difficult to give precise references for many of these definitions and assertions; thus, for example, the theorem on direct products IV,3.4, in the special case of global systems, also follows from 1.49 in Gottschalk and Hedlund (1955). However, several of the non-trivial results seem to be new. Chapters III, V and VI are an exposition of several less elementary results, including the local determinacy theorems, extensions of dynamical systems, and extension of sections. The next following two chapters are concerned with transversal theory and the Poincaré-Bendixson theory for continuous dynamical systems on 2-manifolds. The aim of the remaining chapters, I and IX, is described in their introductory remarks. Except for a critical point existence theorem IX,4.3 (which might easily have been established independently of Chapter IX), these chapters do not directly affect the remaining portions.

At this point it is appropriate to warn the prospective reader as to what the present book does *not* contain. The applied mathematician and the mathematically minded physicist may be disappointed in finding few results directly applicable to concrete problems; the abstract form of the Poincaré symmetry principle II,4.16 is an outstanding exception. Thus e. g. assertions such as the Levinson-Smith theorem on periodic solutions of generalized Liénard equations are not included (and the reader is referred to Sansone and Conti (1956)). Similarly, the obvious generalization of Dragilev's comparison theorem (see Dragilev, 1952) to dynamical systems via the transversal theory of chapter VII is not performed. To recompense for this, interest is focused on many principal questions; some of these have direct bearing on problems in discontinuous automatic control (for example I,5.5 may be treated using V,3).

Secondly, all concepts of a statistical character, and in particular all of ergodic theory, are outside the professed scope of this treatise. The last significant omission is that of stability theory, recurrent motions, etc. The author feels that, on resigning metric spaces as carriers, the correct environment for the development of these would be the theory of local dynamical systems on uniform or proximity spaces, with the (possibly non-uniformly continuous) dynamical system continuous in the topology induced by the uniform or proximity structure; hence this would rather be in the class of further developments.

As concerns the formal structure of the book, the main portion composes of chapters, I to IX, and these are divided into sections, e.g. VI,2 refers to Section 2 of Chapter VI. In each section, the individual items (i.e. assertions, definitions, examples and remarks) are numbered consecutively; these are referred to as, for example, VI,2.10 denoting item 10 in VI,2, and with the chapter number omitted in references local to the same chapter. Occasionally it is useful to perform one further subdivision, numbered as in VI,2.10.2. Figures are numbered consecutively within each chapter, and several of the displayed formulas within each section; neither of these is ever referred to from outside these scopes. Exterior references are in the same style, for example Bourbaki (1951, II, § 1), rather than by page numbers; in some cases this may facilitate reading other editions.

Topological Preliminaries

The purpose of this section is only to recall or specify otherwise current terminology and notation, and is not intended to serve as an exposition of the fundamental concepts. However, there is one exception: due to the author's predilection for sequences and convergence (rather then filters, etc., as standard apparatus in topological reasoning), it became necessary to exhibit the corresponding definitions in detail. Again, most of this is possibly not un-familiar, excepting the concept of a generalized subsequence.

The list of concepts and basic results presented in this section was not intended to be complete, and further terms and theorems are interspersed as required in the following chapters. Furthermore, the criteria for the inclusion of any concept here rather than later were purely subjective.

BASIC NOTATION

Sets will often be denoted by upper-case, their elements by lower case letters. (This rule cannot be enforced consistently, since the elements may be sets on their own; among other exceptions, mappings – and there are sets – will be denoted by either type of letter.) The term 'family' is intended to be synonymous with 'set' (and 'system' or 'collection' is not synonymous). Between sets, inclusion is written as in $A \subset B$, set-difference $A - B$, set-unions and meets as in

$$A \cup B \quad \text{or} \quad \bigcup A_i \quad \text{and} \quad A \cap B \quad \text{or} \quad \bigcap A_i$$

respectively. A set consisting of specified elements $x_1, x_2, \dots, x_n$ is occasionally denoted by $\{x_1, x_2, \dots, x_n\}$; a set containing a unique element x is denoted by (x) rather than $\{x\}$, and termed a singleton; the empty set is $\emptyset$. Given sets A, B, their product $A \times B$ is the set of all pairs (x, y) with $x \in A$ and $y \in B$.

The notation $f : A \to B$ is to stand for the statement that f is a mapping of A into B, i.e. that f is a subset of $B \times A$ such that to each $x \in A$ there is $(y, x) \in f$ for a unique $y \in B$ to be denoted by $f(x)$ or f_x. Occasionally such a map f is also written as

$$\{f_x \mid x \in A\}. \tag{1}$$

Also, in the same situation, for any $X \subset A$ and $Y \subset B$, $f(X)$ denotes the set of all $f(x)$ with $x \in X$, and $f^{-1}(Y)$ the set of all $x \in A$ with $f(x) \in Y$;

$$\text{domain } f = A\,, \quad \text{image } f = f(A)\,;$$

f is termed onto if image $f = B$; f is termed 1–1 if $x \neq x'$ in A implies $f(x) \neq f(x')$, i.e. if each $f^{-1}(y)$ is a singleton or empty. In the latter case, the inverse map is denoted by $f^{-1} : f(A) \to A$.

Since maps are sets, the relation $f \subset g$ between maps f, g is meaningful; f is then called a partialization of g, and g an extension of f. If $f : A \to B$ and $A' \subset A$, then $f \mid A'$ denotes the partialization of f to A', i.e. $f \cap (B \times A')$. If $f : A \to B$ and $g : C \to A$, then the composition of f and g is the map $fg : C \to B$ defined by $fg(x) = f(g(x))$ for all $x \in C$. This situation is also described by saying that commutativity obtains in the diagram

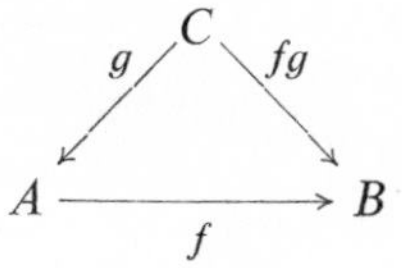

A system of sets $\{X_i \mid i \in I\}$ is said to be centred if

$$\bigcap_{i \in J} X_i \neq \emptyset \quad \text{for all finite} \quad J \subset I\,.$$

The (direct) product of a system of sets $\{P_i \mid i \in I\}$ is denoted by ΠP_i and defined as consisting of all maps $x : I \to \bigcup P_i$ with $x_i \in P_i$ for all $i \in I$. Such an element $x \in \Pi P_i$ is then usually denoted by $[x_i]$ rather than $\{x_i \mid i \in I\}$. For each $j \in I$ one may then define the natural projection map $p_j : \Pi P_i \to P_j$ by

$$p_j[x_i] = x_j\,.$$

Given a set P and an equivalence relation $\sim$ on P (i.e. a reflexive, symmetric and transitive relation), then $P/\sim$ will denote the set of all equivalence classes of P modulo $\sim$; thus $x' \in P/\sim$ if $x' = \{y \in P : y \sim x\}$ for some $x \in x'$. Then there is the natural quotient mapping $e : P \to P/\sim$ defined by taking for $e(x)$ the unique equivalence class containing x. A subset $X \subset P$ is then termed $\sim$saturated if, similarly, $y \sim x \in X$ implies $y \in X$.

If $\mathscr{V}$ is a property significant for elements of a set P, then a notation such as

$$\{x \in P : x \text{ has } \mathscr{V}\}$$

is used to describe the subset of P consisting of all $x \in P$ having property $\mathscr{V}$. This notation may be distinguished from that of (1) by remarking the vertical bar in (1), contrasting with the present colon.

PARTICULAR SETS

C^1 denotes the set of all integers, C^+ its subset consisting of the non-negative integers.

R^1 denotes the set of real numbers, R^+ and R^- the subsets of non-negative and non-positive reals respectively. The set R^1 augmented by two further elements, $+\infty$ and $-\infty$, will be denoted by $\mathsf{R}^\#$; for $x \in \mathsf{R}^1$ one then defines

$$-\infty < x, \quad x < +\infty,$$
$$-\infty + x = -\infty + -\infty = x + -\infty = -\infty,$$
$$+\infty + x = +\infty + +\infty = x + +\infty = +\infty.$$

For x, y in $\mathsf{R}^\#$ let

$$(x, y) = \{t \in \mathsf{R}^\# : x < t < y\}, \quad [x, y] = \{t \in \mathsf{R}^\# : x \leqq t \leqq y\}, \quad \text{etc.}$$

These are called an open-interval and a closed-interval respectively; when necessary, these latter terms will be employed to distinguish between, for example, an open-interval (x, y) and a pair (x, y) as an element of $\mathsf{R}^\# \times \mathsf{R}^\#$.

For positive integral n, Euclidean n-space R^n is defined as the product of n replicas of R^1; its elements may also be described by n-tuples $[x_k]$ $(k = 1, 2, \ldots, n,\ x_k \in \mathsf{R}^1)$. By analogy one also defines $\mathsf{R}^0 = (0)$, a singleton; $\mathsf{R}^{-1} = \emptyset$; and

$$\mathsf{R}^\infty = \{x : x = \{x_k \mid k \in \mathsf{C}^+,\ k > 0\},\ x_k \in \mathsf{R}^1\}.$$

Then R^n may conveniently be identified with the set

$$\{[x_k] \in \mathsf{R}^\infty : x_k = 0 \text{ for } k > n\};$$

and one may define the n-sphere (for integral $n \geqq -1$)

$$\mathsf{S}^n = \{[x_k] \in \mathsf{R}^{n+1} : \textstyle\sum x_k^2 = 1\},$$

the Euclidean n-simplex

$$\mathsf{E}^n = \{[x_k] \in \mathsf{R}^{n+1} : x_k \in \mathsf{R}^+,\ \textstyle\sum x_k = 1\},$$

and the Hilbert parallelepiped

$$\mathsf{E}^\infty = \{[x_k] \in \mathsf{R}^\infty : 0 \leqq x_k \leqq 1,\ \textstyle\sum x_k^2 < +\infty\}.$$

TOPOLOGICAL SPACES

The basic reference here is Bourbaki (1951). A topological space consists of a set P and a topology τ on P; the latter is a family of subsets of P including both $\emptyset$ and P, such that the union of an arbitrary subfamily of τ, and also the meet of any finite subfamily of τ, both belong to τ. A subset $U \subset P$ is termed a neighbourhood of a subset $X \subset P$ (in particular of a singleton $X = (x)$, $x \in P$) if $X \subset G \subset U$ for some $G \in \tau$. The closure of a subset $X \subset P$ is denoted by $\overline{X}$ and consists of all $x \in P$ such that $G \cap X \neq \emptyset$ whenever $x \in G \in \tau$. The interior and the frontier of a subset $X \subset P$ are then defined by

$$\operatorname{Int} X = P - \overline{P - X}\,, \quad \operatorname{Fr} X = \overline{X} \cap \overline{P - X}$$

respectively. Sets belonging to τ are characterized by $X = \operatorname{Int} X$ and termed open; their complements are characterized by $X = \overline{X}$ and termed closed.

Occasionally topologies are defined by the following procedure. There is given a set P and an otherwise arbitrary family $\mathscr{S}$ of subsets of P, the subbasic open sets. One then forms the family $\mathscr{B}$ consisting of all set-meets of finite subfamilies of $\mathscr{S}$, obtaining the basis open sets (this includes $P = \bigcap_{X \in \emptyset} X \in \mathscr{B}$); finally τ is obtained as the family of all unions of arbitrary subsets of $\mathscr{B}$. The natural topology on R^1 may be defined thus, by taking for $\mathscr{S}$ (or $\mathscr{B}$) the family of all the open-intervals.

If (P, τ) is a topological space and the topology τ is given previously or is immaterial, one usually speaks of the topological space P, of subsets open in P, etc.

Let $f : P \to P'$ be a map between topological spaces. Then f is called continuous if $f^{-1}(G')$ is open in P whenever G' is open in P'; f is called interior if $f(G)$ is open in P' whenever G is open in P. If f is $1-1$ onto and is both continuous and interior, then it is called a homeomorphism, and this is indicated by $f : P \approx P'$.

If τ and τ' are topologies on the same set P and $\tau \subset \tau'$, then τ' is called finer than τ, and τ weaker than τ'. The weakest topology of all, $\tau = \{\emptyset, P\}$, is called discrete. Let P be a topological space and Q a subset; then one may define a standard subset topology on Q as the weakest topology rendering continuous the inclusion map

$$i : Q \subset P\,, \quad i(x) = x \quad \text{for} \quad x \in Q\,.$$

It is easily established that $X \subset Q$ is open in the subset topology iff $X = Q \cap G$ with G open in P. If also $r : P \to Q$ is a continuous map with $r(x) = x$ for all $x \in Q$, then r is called a retraction, and Q a retract (of P).

If P_i are topological spaces for $i \in I$, then on the the product ΠP_i one defines the product (Tychonov) topology as the weakest topology rendering continuous all projections $p_j : \Pi P_i \to P_j$ for $j \in I$. If P is a topological space and $\sim$ an equivalence relation on P, then on the corresponding equivalence class set $P/\sim$ one defines the quotient topology as the finest that renders continuous the quotient map $e : P \to P/\sim$. The equivalence relation $\sim$ is then called open if the map e is interior. For further results see Bourbaki (1951). Thus the natural topology on R^1 induces a product topology on R^n and R^∞, and hence subset topologies for S^n, E^n, E^∞, R^+, R^-.

For many purposes it is convenient to classify topological spaces according to their so-called separation properties (Alexandroff and Hopf, 1935). Thus, a topological space P is said to be a

T_0 space (or *Riesz* space) if to any two distinct points in P there exists a neighbourhood of one not containing the other;

T_1 space (or *Kolmogorov* space) if to any two distinct points in P there exist neighbourhoods of either one not containing the other;

T_2 space (or *Hausdorff* space) if any two distinct points in P possess disjoint neighbourhoods;

T_3 space (or *regular* space) if each neighbourhood of any $x \in P$ contains a closed neighbourhood of x;

T_ϱ space (or *Tychonov* or *completely regular* space) if it is T_2 and to each $x \in P$ and any neighbourhood U of x there exists a continuous $f : P \to \mathsf{R}^1$ with $f(x) = 0$ and $f \mid (P - U) = 1$;

T_4 space (or normal space) if it is T_2 and any two disjoint closed subsets of P possess disjoint neighbourhoods;

perfectly normal space if it is T_2 and to any closed subset $F \subset P$ there is a continuous $f : P \to \mathsf{R}^1$ with $F = \{x : f(x) = 0\}$.

COMPACT AND CONNECTED SETS

A topological space P is termed quasi-compact if it satisfies either of the following two equivalent conditions (Bourbaki, 1951):

1. Each centred family of closed subsets of P has non-void intersection.
2. Each open cover of P contains a finite open cover, i.e. if

$$\bigcup_{i \in I} G_i = P \quad \text{with } G_i \text{ open },$$

then also $\bigcup_{i \in J} G_i = P$ for some finite $J \subset I$. Each quasi compact T_2 space is T_4, whereupon it is called compact. One then has the celebrated (Bourbaki, 1948)

Tychonov theorem. *A space P is a subspace of some compact space if P is T_ϱ.*

In the positive case there may even be exhibited a standard compactification βP, in a certain sense maximal.

A topological space P is called connected if

$$P = Q_1 \cup Q_2 , \quad Q_k \text{ closed disjoint },$$

implies $Q_1 = \emptyset$ or $Q_2 = \emptyset$. A general topological space P has a uniquely determined decomposition into components

$$P = \bigcup Q_i$$

with the Q_i disjoint, closed and connected, and such that any connected subset of P is entirely contained within some Q_i. It should be remarked that a subset Q of a topological space P is called connected (or compact, or pathwise connected, etc.) if it is such in the subset topology inherited from P.

A topological space P is called pathwise connected if to any x, y in P there is a continuous $f:[0, 1] \to P$ with $f(0) = x$, $f(1) = y$; in any case such an f is termed a path in P, from x to y, and occasionally is denoted by $\widehat{xy}$. A topological space is called locally quasi-compact (or locally connected or locally pathwise connected) if every neighbourhood of any $x \in P$ contains a quasi-compact (or connected, or pathwise connected, respectively) neighbourhood of x. A locally quasi-compact T_2 space is called a locally compact or LC space; a T_2 space is LC if each point has a compact neighbourhood. For further properties of these types of space see Kuratowski (1950).

MANIFOLDS

A topological space P is called a (topological) n-manifold if it is a connected T_2 space and each point has a neighbourhood homeomorphic to R^n ($n \geqq -1$ a given integer). If in this definition one replaces R^n by E^n, the resulting object is called an n-manifold with boundary (to be used as an indivisible term); the boundary then consists of all points which do not have a neighbourhood homeomorphic to R^n. It may be noted that each such space is LC and locally pathwise connected. The following celebrated result will often be applied (Eilenberg and Steenrod, 1952).

Preservation of domain theorem. *If X and Y are homeomorphic subsets of n-manifolds and X is open, then so is Y.*

There follows immediately (see Alexandroff, 1947): If $f: X \approx Y$ and both X, Y are n-manifolds with boundary, then f maps the boundary of X onto that of Y.

In I,3 there occur differential manifolds; in describing these it is first necessary to introduce differential coordinate systems (Steenrod, 1951). Let P be an n-manifold as above; a coordinate system (differential, of class 1) on P is a set C of mappings with the following properties. Each $c \in C$ is a homeomorphism of an open subset of R^n into P; for $c \in C$ the sets *image* c constitute an (open) cover of P; for any c, c' in C, the map $c^{-1}c'$ is differentiable on c'^{-1} (*image* c). Two coordinate systems C and C' on P are called equivalent if $C \cup C'$ is again a coordinate system; this defines an equivalence relation, and each equivalence class is called a differential n-manifold. For further allied concepts, in particular orientability, see *loc. cit.*

GENERALIZED SEQUENCES

A directed set consists of a non-void set I, and of a transitive relation $\geqq$ on I (the order relation) with the Moore-Smith "roofing" property that to any i_1, i_2 in I

there is a common $j \in I$ such that

$$j \geqq i_1\,, \quad j \geqq i_2$$

(Birkhoff, 1948). A subset $J \subset I$ is called confinal to $(I, \geqq)$ if to any $i \in I$ one has $j \geqq i$ for some $j \in J$. It may be noted then that J together with $\geqq$ partialized over J (i.e. partialized to $J \times J$) constitutes a directed set on its own; however, no relation on J intervenes in the definition itself. As examples, C^+ is confinal to R^1 ordered naturally; a subset of C^+ is confinal to C^+ iff it is infinite.

In speaking about directed sets $(I, \geqq)$, the second term, $\geqq$, is often omitted as being understood implicitly; these and other useful but lax formulations such as "the order relation of the directed set I" should then be regarded in the nature of abbreviations.

A (generalized) sequence is any map of a directed set, termed the set of indices of the sequence. Map values will then be denoted by indexing; a preferred notation for sequences is then $\{x_i \mid i \in I\}$ (or $\{x_i\}$, or $\{x_n\}_{n=0}^{\infty}$ for C^+ as set of indices, etc.), or even clauses such as "the sequence x_i".

Given two sequences

$$\{x_i \mid i \in I\}\,, \quad \{y_j \mid j \in J\}\,, \tag{2}$$

the sequence y_j is said to be confinal to x_i (*loc. cit.*) if J is a confinal subset of I, the order relation of J coincides with that of I partialized over J, and finally

$$y_j = x_j \quad \text{for all} \quad j \in J\,.$$

Evidently all sequences confinal to a given one $\{x_i \mid i \in I\}$ may be obtained simply by partializing the assignment $i \to x_i$ to some confinal subset of the set I of indices.

It will also be necessary to introduce a considerably less restrictive concept. In (2), the sequence y_j will be termed a subsequence of the sequence x_i iff there exist maps $f, g : J \to I$ with the following properties:

$$y_j = x_{f(j)}\,, \quad f(j) \geqq g(j) \quad \text{for all} \quad j \in J\,,$$

g maps onto a confinal subset $g(J)$ of I, and g is order-preserving (i.e. $i \geqq j$ in J implies $g(i) \geqq g(j)$ in I).

As will be shown below, this definition is useful; however, as it is a little involved it may be more useful to exhibit an example in a more familiar context. For $I = J$ take C^+ with the natural order; let there be given a sequence $\{x_i \mid i \in \mathsf{C}^+\}$ and also a map $f : \mathsf{C}^+ \to \mathsf{C}^+$. Form the sequence $\{x_{f(i)} \mid i \in \mathsf{C}^+\}$ and consider whether or not it is a subsequence of the x_i; by definition, this question reduces to inquiring about the existence of the map g with the indicated properties. To this end define

$$g(i) = \min_{j \geqq i} f(j)\,,$$

so that g is the maximal non-decreasing (i.e. order-preserving) map $\mathsf{C}^+ \to \mathsf{C}^+$ majorized by f. Thus $x_{f(i)}$ is a subsequence of x_i iff g maps onto a confinal subset of C^+, i.e. iff $g(\mathsf{C}^+)$ is infinite. It is easily seen that this obtains iff, for each $n \in \mathsf{C}^+$, the set

$$\{i \in \mathsf{C}^+ : f(i) \leqq n\}$$

is finite, i.e. iff $f(i) \to +\infty$. Notice that, in the same situation, $x_{f(i)}$ is a confinal sequence to x_i iff f is strictly increasing.

It may be remarked that a sequence confinal to x_i is necessarily a subsequence of x_i: for $f = g$ merely take the inclusion map. Furthermore, the relation of being a subsequence is transitive: if y_j is a subsequence of x_i relative to $f, g : J \to I$ as above, and similarly z_k is a subsequence of y_j relative to $f', g' : K \to J$, then z_k is a subsequence of x_i relative to ff', $gg' : K \to I$. In establishing this assertion observe that

$$z_k = y_{f'(k)} = x_{ff'(k)} \quad \text{for} \quad k \in K\,,$$

that $gg'(K)$ is confinal to $g(J)$ and hence to I, and finally that the assumptions on the maps concerned yield

$$ff'(k) \geqq gf'(k) \geqq gg'(k) \quad \text{for} \quad k \in K$$

with gg' order-preserving.

CONVERGENCE

From now on assume given a topological space P. A fundamental example of a directed set is afforded by the system $\mathscr{V}_x$ of all neighbourhoods of some fixed $x \in P$, ordered by set-inclusion:

$$U \geqq V \quad \text{iff} \quad U \subset V\,.$$

A sequence $\{x_i \mid i \in I\}$ in P is said to converge to a point $x \in P$ as its limit, and this is denoted by

$$x_i \to x \quad \text{or} \quad \lim x_i \ni x\,,$$

if to every neighbourhood U of x there is an index $j \in I$ with $x_i \in U$ for all $i \geqq j$. Some properties are immediate; however, note that a sequence may have none, one, or several limits. If $P \subset P'$ topologically and x_i, $x \in P$, then $x_i \to x$ in P iff $x_i \to x$ in P'. A constant sequence $x_i \equiv x$ converges to x.

Lemma. Each subsequence of a convergent sequence converges to the same limit.

Proof. Let $x_i \to x$ and let y_j be a subsequence of x_i relative to $f, g : J \to I$ as in the definition. Take any neighbourhood U of x. Since $x_i \to x$, there is an index $i' \in I$ with $x_i \in U$ whenever $i \geqq i'$ in I. As $g(J)$ is confinal to I, one may take $j' \in J$ with $g(j') \geqq i'$. For all $j \geqq j'$ in J one then has

$$f(j) \geqq g(j) \geqq g(j') \geqq i'\,,$$

so that

$$y_j = x_{f(j)} \in U$$

as required for convergence of y_j.

Further properties of convergence are easily established. Thus, if a sequence x_i does not converge to x, some sequence confinal to x_i has the property that none of its confinal sequences (indeed, no subsequences) converge to x. If I and J are directed sets and $x_{i,j}$ a sequence with the direct product $I \times J$ as index set, and if

$$x_{i,j} \to x_j \quad \text{for fixed} \quad j \in J\,, \quad x_j \to x\,,$$

then some subsequence of $\{x_{i,j} \mid (i, j) \in I \times J\}$ converges to x.

The fundamental connection between topology and convergence is the following easily obtained assertion, *loc. cit.*: If X is any subset of a topological space P, then $x \in \overline{X}$ iff $x_i \to x$ for some sequence x_i in X. Thus the closure of X consists of limits of convergent sequences in X; hence the topology of P itself may be reconstructed from the convergence properties of sequences in P.

Principally, each topological property can be described in terms of convergence. Thus, a topological space P is T_1 iff constant sequences have unique limits; and P is T_2 iff each sequence has at most one limit (i.e. if limits are determined uniquely). In particular, if $x_i \to x$ in a T_2 space and $y_j \to y$ for a subsequence y_j of x_i, then $x = y$. More or less analogous descriptions may be given of T_0 and T_3 spaces. As concerns compactness, one has the following important lemma.

Lemma. A topological space P is quasi-compact iff every sequence in P has a convergent subsequence.

Proof. First let P be quasi-compact and $\{x_i \mid i \in I\}$ any sequence in P. For each $i \in I$ set $X_i = \{x_j : j \geqq i\}$, a subset of P. From the roofing property of I it follows that $\{\overline{X}_i \mid i \in I\}$ is a centred system of non-void closed sets in P; hence there exists a point x,

$$x \in \bigcap_{i \in I} \overline{X}_i\,.$$

To any $i \in I$ and neighbourhood $U \in \mathscr{V}_x$ of x one has that U intersects X_i; using the axiom of choice, for each $i \in I$, $U \in \mathscr{V}_x$, select an index $j(i, U) \in I$ with

$$x_{j(i,U)} \in U\,, \quad j(i, U) \geqq i\,.$$

Now consider the directed set $I \times \mathscr{V}_x$ with the natural direct product order relation

$$(i, U) \geqq (j, V) \quad \text{iff} \quad i \geqq j \quad \text{in } I, U \subset V;$$

and also the order-preserving projection $p : I \times \mathscr{V}_x \to I$ defined by $p(i, U) = i$. Then

$$y_{i,U} = x_{j(i,U)}$$

defines a sequence with index set $I \times \mathscr{V}_x$; obviously $y_{i,U}$ converges to x, and is a subsequence of x_i relative to the maps $j, p : I \times \mathscr{V}_x \to I$; in particular,

$$j(i, U) \geqq i = p(i, U) \quad \text{in} \quad I \times \mathscr{V}_x .$$

For the converse assertion let the described situation obtain, and take any centred family $\mathscr{F}$ of closed subsets of P. Under set-inclusion, the family of finite subsets of $\mathscr{F}$ constitutes a directed set I; and by assumption, for each

$$i = \{F_1, F_2, \dots, F_n\} \in I , \quad F_k \in \mathscr{F} ,$$

one may choose a point $x_i \in \bigcap F_k$. From the given condition, there exists a convergent sequence $y_j \to y$ which is a subsequence of x_i relative to maps say $f, g : J \to I$. To prove that P is quasi-compact it now suffices to verify that y is in all $F \in \mathscr{F}$.

Thus, take any $F \in \mathscr{F}$; then take the singleton $\{F\} \in I$, so that there is a $j \in J$ with $g(j) \geqq \{F\}$. Also take any neighbourhood U of y, so that there is a $j' \in J$ with $y_i \in U$ whenever $i \geqq j'$ in J. For $i \geqq j, j'$ one then has simultaneously,

$$f(i) \geqq g(i) \geqq g(j) \geqq \{F\} , \quad \text{i.e.} \quad F \in f(i) ,$$
$$x_{f(i)} = y_i \in U .$$

Therefore $x_{f(i)} \in F$, and U intersects F; since this holds for all neighbourhoods U of y, it follows that $y \in F$ as required. This completes the proof.

Within a topological space P, one may define limits of sequences of subsets $X_i \subset P$, as follows. Let $\operatorname{liminf} X_i$ consist of all limits of sequences of points $x_i \in X_i$, in the sense described previously; let $\operatorname{limsup} X_i$ consist of all limits of sequences of points $y_j \in Y_j$ with Y_j a subsequence of X_i. Obviously then

$$\operatorname{liminf} X_i \subset \operatorname{limsup} X_i ;$$

and if equality obtains in this relation, the sequence X_i is said to converge to its limit,

$$\lim X_i = \operatorname{liminf} X_i = \operatorname{limsup} X_i .$$

It is easily shown that if the sequence X_i has the property that $X_i \supset \overline{X}_j$ whenever $i \leqq j$, then

$$\lim X_i = \bigcap X_i .$$

CATEGORIES

Here an authoritative text is Kurosh (1960). Roughly speaking, a category $\mathscr{C}$ consists of the so-called objects and morphisms of $\mathscr{C}$, and of the composition in $\mathscr{C}$, which is a partial binary operator on the morphisms. The requirements are as follows. The morphisms are grouped into disjoint sets $\mathcal{M}_{\mathscr{C}}(X, Y)$ for any objects X, Y in $\mathscr{C}$ (and $f \in \mathcal{M}_{\mathscr{C}}(X, Y)$ is also written as $f : X \to Y$ in $\mathscr{C}$). If

$$f : X \to Y \text{ in } \mathscr{C}, \quad f' : X' \to Y' \text{ in } \mathscr{C},$$

then the composition ff' of f and f' is defined iff $X = Y'$, whereupon $ff' : X' \to Y$ in $\mathscr{C}$. It is required that composition be associative, $f_1(f_2 f_3) = (f_1 f_2) f_3$ whenever either side is defined; and also that in each $\mathcal{M}_{\mathscr{C}}(X, X)$ there exist a morphism i such that

$$fi = f, \quad ig = g \tag{3}$$

whenever the left sides are defined.

It is then easily shown that (3) determine $i : X \to X$ uniquely, so that to each object X in $\mathscr{C}$ there corresponds, in a $1-1$ manner, the morphism i called the identity morphism of X; this is denoted by $i : X = X$ in $\mathscr{C}$. Essentially, then, the objects of a category $\mathscr{C}$ are entirely superfluous, serving only to label the identities of $\mathscr{C}$.

As an example consider the category described as follows. The objects of $\mathscr{S}$ are all the sets; the morphisms in $\mathcal{M}_{\mathscr{S}}(X, Y)$ are all pairs (f, Y) where f is a mapping of X into Y; and the composition of

$$(f, Y) : X \to Y, \quad (f', X) : X' \to X$$

is defined as (ff', Y) with ff' denoting the set-theoretical composition of the maps f and f'. (The point of having the second term in these pairs (f, Y) is to ensure that the $\mathcal{M}_{\mathscr{S}}(X, Y)$ be disjoint.)

Any subcategory of $\mathscr{S}$ (a concept defined in the possibly obvious manner) is called a concrete category; all the categories appearing in the present book are concrete.

Given two categories $\mathscr{C}$ and $\mathscr{C}'$, a (covariant) functor $F : \mathscr{C} \to \mathscr{C}'$ is a mapping of the morphisms of $\mathscr{C}$ into those of $\mathscr{C}'$ which sends identities into identities and has

$$F(f_1 f_2) = F(f_1)\, F(f_2)$$

for all morphisms f_k in $\mathscr{C}$. Necessarily then F associates with each object X in $\mathscr{C}$ an object X' in $\mathscr{C}'$, namely that corresponding to the identity $F(i)$ having $i : X = X$ in $\mathscr{C}$; denoting X' by $F(X)$, one then has that

$$f : X \to Y \text{ in } \mathscr{C} \quad \text{implies} \quad F(f) : F(X) \to F(Y) \text{ in } \mathscr{C}'.$$

CHAPTER I

Differential Equations

The main purpose of this chapter is to introduce the subjects studied in the following chapters, mainly by exhibiting a sufficiently familiar concrete model for the more abstract constructions to be met with later. Furthermore, it is intended to suggest that the degree of generality adopted is sufficiently motivated by current but non-classical problems in differential equation theory, and in branches of physical science which use differential equations as a significant tool. Section 5 in particular aims to emphasize the importance of abstract dynamical system theory on 2-manifolds, or rather on covering spaces of the Euclidean plane (also see Neymark 1961).

A reader familiar with differential equations or willing to take the motivation for granted will lose little by passing over this chapter, since the following chapters are, as far as the formal exposition is concerned, entirely independent. Intentionally, most of the references are to survey papers or standard texts.

1. DIFFERENTIAL EQUATIONS

1.1. We shall be much concerned with *differential equations*, more precisely, with first-order ordinary differential equations in Euclidean n-space. Such an equation is usually written as

$$\frac{\mathrm{d}x}{\mathrm{d}\theta} = f(x, \theta) , \tag{1}$$

with $f : D \to \mathrm{R}^n$ a continuous mapping of an open set $D \subset \mathrm{R}^n \times \mathrm{R}^1$; of course (1) corresponds, in current terminology, to a system of n ordinary first-order differential equations. A *solution* of (1) is a map $x : I \to \mathrm{R}^n$ of an open interval $I \subset \mathrm{R}^1$, which "satisfies" (1) in the sense that

$$\frac{\mathrm{d}}{\mathrm{d}\theta} x(\theta) = f(x(\theta), \theta) \quad \text{for} \quad \theta \in I ; \tag{2}$$

or expressly, for all $\theta \in I$, $(x(\theta), \theta) \in D$, $x(\theta)$ is differentiable, and (2) obtains. Sometimes (1) is called a differential system, D the set of states, with solutions called motions, and the independent variable in (1) called time; physical connotations are suggested, but not implied.

In the strict sense, the differential equation is completely determined by the map f as above; all the other symbols appearing in (1) are entirely redundant.

1.2. Recall that mappings are particular types of sets; for example, the solution x above is a special subset of $\mathsf{R}^n \times \mathsf{R}^1$ (such that $(x_0, \theta_0) \in x$ iff $x_0 = x(\theta_0)$). In particular, set-inclusion between maps is well defined, and corresponds to formation of partial maps or extended maps. From the definition of solutions it follows that the set-union of a monotone family of solutions is itself a solution (monotone in the sense that an inclusion obtains between any two members of the family). Applying the Zorn lemma, it results that each solution of (1) is contaired in some maximal or *characteristic* solution; and occasionally the term solution is reserved for these characteristic solutions only.

Concerning existence of solutions of (1) there is the well known

1.3. Cauchy-Peano Theorem. Assume given an open set $D \subset \mathsf{R}^n \times \mathsf{R}^1$ and a continuous map $f : D \to \mathsf{R}^n$. Then to any $(x_0, \theta_0) \in D$, there exists an open interval I and a continuous map $x : I \to \mathsf{R}^n$, such that $\theta_0 \in I$,

$$x(\theta_0) = x_0 , \tag{3}$$

$$\frac{\mathrm{d}}{\mathrm{d}\theta} x(\theta) = f(x(\theta), \theta) \quad \textit{for} \quad \theta \in I .$$

Proofs of 1.3 may be found in current text-books; an elegant one appears in Franklin, 1947. In the situation described, (x_0, θ_0) are called initial data, the relation (3) is called an *initial condition*, and (1), (3) combined are called an initial value problem.

1.4. If x_1 is a characteristic solution of (1) and (β, α) its domain, then it cannot happen that both

$$\lim_{\theta \to \alpha -} x_1(\theta) = x_0$$

and also $(x_0, \alpha) \in D$ exist. In the opposite case, using 1.3 one would also have a second solution x_2 to satisfy initial data (x_0, α) and defined on some $(\alpha - \varepsilon, \alpha + \varepsilon)$, whereupon the map x described by

$$x(\theta) = \begin{cases} x_1(\theta) & \text{for} \quad \beta < \theta < \alpha \\ x_2(\theta) & \text{for} \quad \alpha \leqq \theta < \alpha + \varepsilon \end{cases}$$

would be a solution and a proper extension of the maximal solution x_1.

1.5. As all good theorems do, 1.3 raises further natural questions; at present we shall state these quite informally:

1. Is a characteristic solution of (1) determined uniquely by specifying the initial data?

2. Do small changes in initial data evoke small variations in the corresponding solutions?

3. If the map f which determines (1) is varied slightly, do the corresponding solutions also change little?

Of course, 2 and 3 may be combined to form an even more general problem. It is known that 2 is a special case of 3, and that 1 and 2 are not unconnected.

1.6. To obtain meaningful answers, it is necessary to interpret the terms used in 1–3 in some definite manner; first consider 1.

Given f and D = *domain* f as in 1.1, a point $(x_0, \theta_0) \in D$ is termed a *point of unicity* of (1) if, to any pair of solutions x_1, x_2 of (1) with

$$x_1(\theta_0) = x_0 = x_2(\theta_0),$$

there is an $\varepsilon > 0$ such that

$$x_1(\theta) = x_2(\theta) \quad \text{if} \quad |\theta - \theta_0| < \varepsilon .$$

If D consists exclusively of points of unicity, then (1) is said to have *unicity of solutions*; in particular, a simple argument shows that each characteristic solution x is then uniquely determined by initial data (x_0, θ_0) with $x_0 = x(\theta_0)$.

In general, equation (1) does not have unicity of solutions; a familiar example is the differential equation in R^1

$$\frac{dx}{d\theta} = (x^2)^{1/3}$$

[$D = \mathrm{R}^2$; solutions analytic in θ:

$$x(\theta) = \tfrac{1}{27}(\theta - \theta_0)^3 \quad \text{for fixed} \quad \theta_0 \in \mathrm{R}^1 ,$$
$$x(\theta) \equiv 0.]$$

A rough-and-ready unicity condition for (1) is continuity of the jacobian matrix $\partial f/\partial x$ in D; for example, if the differential equation (1) is linear (as the term is used in differential equation theory), i.e.

$$f_i(x, \theta) = \sum_{j=1}^{n} \varphi_{ij}(\theta)\, x_j + \varphi_i(\theta) \tag{4}$$

with the f_i as coordinates of f, and x_i as coordinates of x, then $\partial f/\partial x$ is the matrix (φ_{ij}). Considerable effort has been devoted to obtaining more and more refined conditions ensuring unicity. One of these is presented below (Sansone and Conti, 1956, I, § 3.1); here and elsewhere (x, y) and $\|x\|$ denote the usual Euclidean scalar product and norm in n-space.

1.7. Theorem. Let $f : D \to \mathrm{R}^n$ be continuous, $D \subset \mathrm{R}^n \times \mathrm{R}^1$ open, $(x_0, \theta_0) \in D$, and consider (1). Assume that there exist:

1. an open neighbourhood $G \times I$ with $(x_0, \theta_0) \in G \times I \subset D$;

2. a continuous positive function $\varrho : (0, \varepsilon) \to \mathsf{R}^1$ *with*

$$\int_0^\varepsilon \frac{d\theta}{\varrho(\theta)} = +\infty ;$$

3. locally Lebesgue summable functions σ, τ *with*

$$\|f(x, \theta)\| \leqq \tau(\theta) \quad \text{for} \quad x \in G\,, \quad \theta \in I\,,$$
$$(f(x, \theta) - f(y, \theta), x - y) \leqq \sigma(\theta)\, \varrho(\|x - y\|^2) \quad \text{for} \quad x, y \in G$$

and $\theta \in I$.

Then (x_0, θ_0) *is a point of unicity of (1).*

1.8. As concerns the problems 1.5.2 and 3, any formalization must specify what is meant by slight variations of a mapping; thus it must introduce, directly or not, some continuity structure on the set of admissible solutions and maps f.

A quite reasonable topology, but by no means the only reasonable one, is that of almost uniform (or compact) convergence for maps on a common domain $X \subset \mathsf{R}^m$ into R^n: let $f_i \to f$ almost uniformly in X mean that, for each compact $F \subset X$, f_i converges to f uniformly on F. As evidence of reasonability of this topology, one has that if $f : G \times A \to \mathsf{R}^n$ is continuous, $G \times A \subset \mathsf{R}^p \times \mathsf{R}^q$, then

$$f(x, \lambda) \to f(x, \lambda_0) \quad \text{as} \quad \lambda \to \lambda_0 \text{ in } A$$

almost uniformly for $x \in G$. The following theorem concerns precisely this situation (Coddington and Levinson, 1955, II, 4.3):

1.9. Theorem. Let $D \subset \mathsf{R}^n \times \mathsf{R}^1$ *be open,* $I \subset \mathsf{R}^1$ *an open interval,* $f : D \times \times I \to \mathsf{R}^n$ *a bounded continuous map, and for each* $(x', \theta', \lambda) \in D \times I$ *consider the initial value problem*

$$\frac{dx}{d\theta} = f(x, \theta, \lambda)\,, \tag{5}$$
$$x(\theta') = x'\,.$$

Take an $(x_0, \theta_0, \lambda_0) \in D \times I$, *and assume that (5) with* $(x', \theta', \lambda) = (x_0, \theta_0, \lambda_0)$ *has a unique solution* $x : [\beta, \alpha] \to \mathsf{R}^n$ *with* $\beta < \theta_0 < \alpha$.

Then, for each $C = (x', \theta', \lambda) \in D \times I$ *sufficiently near to* $(x_0, \theta_0, \lambda_0)$, *(5) has a solution* $x_C : [\beta, \alpha] \to \mathsf{R}^n$ *such that*

$$x_C \to x \quad \textit{uniformly on } [\beta, \alpha]$$

as $C = (x', \theta', \lambda) \to (x_0, \theta_0, \lambda_0)$.

1.10. Remarks. The boundedness condition on f is entirely irrelevant; as concerns the unicity assumption on x, it may also be omitted, on modifying the conclusions of *1.9* appropriately (cf. Sansone, 1949, p. 82; Hájek 1, 1960, 1.6).

Note also the following consequence: for C near $(x_0, \theta_0, \lambda_0)$, the solution x_C must be defined on $[\beta, \alpha]$; hence if α_C denotes the upper bound of the interval domain of the corresponding characteristic solution, one must have

$$\alpha \leqq \liminf \alpha_C ;$$

and, of course, analogously for the lower bound.

1.11. Consider the special case of (1) in which the set $D = \textit{domain} f$ has the special form

$$D = G \times \mathsf{R}^1 ,$$

so that G is open in R^n; and let (β, α) be the interval domain of a characteristic solution x of (1). It may happen that $\alpha = +\infty$, and then x is termed extendable into the future; in the opposite case $0 < \alpha < +\infty$, x is said to have finite positive *escape time* α. Similar terms are applied to the lower bound β. If all characteristic solutions of (1) are extendable into both the past and the future, then (1) is said to have ***extendability of solutions.***

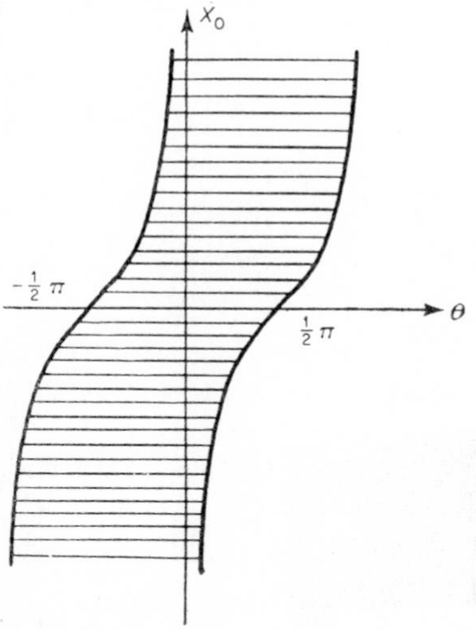

Fig. 1. Domain T in dependence on x_0.

A well known instance of finite escape times is the equation in R^1

$$\frac{dx}{d\theta} = x^2 + 1 ;$$

unicity of solutions obtains, and the characteristic solution with initial data $(x_0, \theta_0) \in \mathsf{R}^1 \times \mathsf{R}^1$ is described by

$$x(\theta) = \tan(\theta - \theta_0 + \arctan x_0) ,$$

with arctan interpreted in the strict sense as the unique function inverse to tan restricted to $(-\frac{1}{2}\pi, \frac{1}{2}\pi)$ as domain. The domain of this solution x is the open interval

$$(-\tfrac{1}{2}\pi + \theta_0 - \arctan x_0 , \quad \tfrac{1}{2}\pi + \theta_0 - \arctan x_0) ,$$

cf. Fig. 1 for $\theta_0 = 0$.

Again, there is an easily applicable condition for extendability of solutions in (1) with $D = \mathsf{R}^n \times \mathsf{R}^1$: that there exist a constant σ such that

$$\|f(x, \theta)\| \leqq \sigma(1 + \|x\|) \quad \text{for all} \quad x \in \mathsf{R}^n , \quad \theta \in \mathsf{R}^1 .$$

Thus, in the case of a linear equation as in (4), uniform boundedness of all φ_{ij}, φ_i is sufficient for extendability.

A more refined condition is given below (see Sansone and Conti, 1956, I, § 1.3, which cites further references; also Lassalle and Lefschetz, 1957 § 23).

1.12. Theorem. Let $D = \mathsf{R}^n \times (\beta, \alpha)$ *with* $\alpha = +\infty$ *allowed,* $f : D \to \mathsf{R}^n$ *be continuous,* $(x_0, \theta_0) \in D$, *and consider the initial value problem*

$$\frac{\mathrm{d}x}{\mathrm{d}\theta} = f(x, \theta), \quad x(\theta_0) = x_0 . \tag{6}$$

Assume that there exist

1. a continuous positive function $\varrho : (0, +\infty) \to \mathsf{R}^1$ *with*

$$\int^{+\infty} \frac{\mathrm{d}\theta}{\varrho(\theta)} = +\infty ;$$

2. a continuous function $\sigma : [\theta_0, \alpha) \to \mathsf{R}^1$ *such that*

$$\|f(x, \theta)\| \leqq \sigma(\theta)\, \varrho(\|x\|) \quad \textit{for} \quad x \in \mathsf{R}^n,\ \theta \in [\theta_0, \alpha) ;$$

then each characteristic solution of (6) *is defined on* $[\theta_0, \alpha)$.

(*Remark.* The formulation *loc. cit.* concerns another norm in R^n, but at least equivalent to the Euclidean norm $\|x\|$.)

1.13. The following remarks may well appear, in the present context, both trivial and unmotivated; however they will constitute the basis of far-reaching generalizations in Chapters II and IX.

Assume given a differential equation (1) with *domain* $f = G \times \mathsf{R}^1$ open in $\mathsf{R}^n \times \mathsf{R}^1$, and assume both unicity and extendability of solutions. Take any α, β, γ in R^1. With each $x_0 \in G$ perform the following procedure. First determine the (characteristic) solution x to the initial condition

$$x(\gamma) = x_0 ;$$

having x, determine the solution y to the initial condition

$$y(\beta) = x(\beta) ; \tag{7}$$

finally, consider the values of x and y at $\theta = \alpha$. From unicity and (7) it follows that $x(\theta) \equiv y(\theta)$, so that, in particular, $x(\alpha) = y(\alpha)$.

This may be formulated in another manner, possibly unnecessarily sophisticated. For any $\alpha, \beta \in \mathsf{R}^1$ define a map ${}_\alpha T_\beta : G \to G$; thus to any $x_0 \in G$ find the solution x with initial condition

$$x(\beta) = x_0 ,$$

and then set

$$_{\alpha}T_{\beta}\, x_0 = x(\alpha)\,.$$

This may be expressed vaguely by saying that the mapping $_{\alpha}T_{\beta}$ describes movement along the solutions. Obviously $_{\alpha}T_{\alpha}$ is the identity map of G, and

$$_{\alpha}T_{\beta}\, {}_{\beta}T_{\gamma} = {}_{\alpha}T_{\gamma} \tag{8}$$

represents the result obtained above.

Next, notice the special case of equation (1) when the map f does not depend on θ.

1.14. Proposition. Let $f : G \times \mathsf{R}^1 \to \mathsf{R}^n$ *be continuous,* G *open in* R^n, *and consider the differential equation*

$$\frac{\mathrm{d}x}{\mathrm{d}\theta} = f(x, \theta)\,. \tag{1}$$

Then f *is independent of* θ *iff, for any solution* $x : (\beta, \alpha) \to \mathsf{R}^n$ *of (1) and each* $\theta_0 \in (\beta, \alpha)$, *the map* y *defined by*

$$y(\theta) = x(\theta + \theta_0) \quad \text{for} \quad \beta - \theta_0 < \theta < \alpha - \theta_0 \tag{9}$$

is again a solution of (1).

Proof. This is virtually straightforward verification. For an f independent of θ, write $f(x, \theta) = g(x)$; then x satisfies

$$\frac{\mathrm{d}}{\mathrm{d}\theta}\, x(\theta) = g(x(\theta))\,,$$

so that

$$\frac{\mathrm{d}}{\mathrm{d}\theta}\, y(\theta) = \frac{\mathrm{d}}{\mathrm{d}\theta}\, x(\theta + \theta_0) = g(x(\theta + \theta_0)) = g(y(\theta))\,,$$

and y is indeed a solution.

Conversely, take any $(x_0, \theta_0) \in G \times \mathsf{R}^1$; from 1.3, there exists a solution x of (1) to the initial condition $x(\theta_0) = x_0$, and whose domain (β, α) has $\beta < \theta_0 < \alpha$. Hence, for $\theta = \theta_0$,

$$\frac{\mathrm{d}}{\mathrm{d}\theta}\, x(\theta)\big|_{\theta = \theta_0} = f(x(\theta_0), \theta_0) = f(x_0, \theta_0)\,. \tag{10}$$

By assumption, the mapping y of (9) is again a solution of (1), so that

$$\frac{\mathrm{d}}{\mathrm{d}\theta}\, x(\theta + \theta_0) = \frac{\mathrm{d}}{\mathrm{d}\theta}\, y(\theta) = f(y(\theta), \theta) = f(x(\theta + \theta_0), \theta)\,.$$

Setting $\theta = 0$ and comparing with (10) we obtain

$$f(x_0, \theta_0) = f(x_0, O)\,.$$

Since (x_0, θ_0) was an arbitrary point in $G \times \mathsf{R}^1$, this shows that f is indeed independent of θ.

1.15. If the differential equation (1) describes the behaviour of a physical system in time, θ, then the case examined in 1.14 corresponds to that where the physical law governing the behaviour in question is time-independent. Such systems are called *autonomous*, and this term will also be used for the differential equations.

A typical instance is that of a system consisting of a single particle of unit mass, moving in a conservative field of force. The Newtonian equations of motions are (with vector notation in R^3)

$$\ddot{x} = f(x)\,;$$

these correspond to an equation in the form of (1) in $\mathsf{R}^3 \times \mathsf{R}^3$

$$\frac{\mathrm{d}x}{\mathrm{d}\theta} = y\,, \quad \frac{\mathrm{d}y}{\mathrm{d}\theta} = f(x)\,.$$

1.16. Under the assumptions of 1.13, i.e. unicity and extendability, another necessary and sufficient condition may be obtained in 1.14: for all $\alpha, \beta \in \mathsf{R}^1$,

$$_{\alpha}T_{\beta} = {}_{\alpha-\beta}T_0\,. \tag{11}$$

Indeed, take any $x_0 \in G$. In any case, for some solutions x, y,

$$\begin{aligned} _{\alpha}T_{\beta}\, x_0 &= x(\alpha)\,, & _{\alpha-\beta}T_0\, x_0 &= y(\alpha - \beta)\,,\\ x_0 &= x(\beta)\,, & x_0 &= y(0)\,. \end{aligned}$$

If f is independent of θ then $x(\theta + \beta)$ is a solution of (1) and, from unicity,

$$x(\theta + \beta) = y(\theta) \quad \text{for all} \quad \theta\,;$$

in particular

$$_{\alpha}T_{\beta}\, x_0 = x(\alpha) = x(\alpha - \beta + \beta) = y(\alpha - \beta) = {}_{\alpha-\beta}T_0\, x_0\,,$$

proving (11). Conversely, if (11) holds for all α, β, then

$$x(\theta + \beta) = {}_{\theta+\beta}T_{\beta}\, x_0 = {}_{\theta}T_0\, x_0 = y(\theta)\,,$$

so that $x(\theta + \beta)$ is a solution if $x(\theta)$ is.

1.17. Remarks. From (11) and (8) there follows an important property of the maps $_{\theta}T_0$: for autonomous systems,

$$_{\theta'}T_0\, {}_{\theta}T_0 = {}_{\theta'+\theta}T_0\,. \tag{12}$$

Indeed, from (11),

$$_{\theta'}T_0 = {}_{\theta'+\theta}T_\theta\,,$$

so that

$$_{\theta'}T_0\,{}_\theta T_0 = {}_{\theta'+\theta}T_\theta\,{}_\theta T_0 = {}_{\theta'+\theta}T_0\,.$$

In the general case of equation (1), the equation is completely determined by its solutions, and hence also by the system of maps $\{_\alpha T_\beta \mid \alpha, \beta \in \mathsf{R}^1\}$; from (11) it then follows that an autonomous differential equation is completely determined by the sub system $\{_\theta T_0 \mid \theta \in \mathsf{R}^1\}$.

2. SOME EXAMPLES

2.1. The most general form of a linear autonomous equation in R^2 is, in real coordinates x, y,

$$\frac{\mathrm{d}x}{\mathrm{d}\theta} = a_{11}x + a_{12}y + b_1\,, \quad \frac{\mathrm{d}y}{\mathrm{d}\theta} = a_{21}x + a_{22}y + b_2\,,$$

with the a_{ij}, b_i real and constant. Usually the complex form is more amenable; with $z = x + iy$, one obtains

$$\frac{\mathrm{d}z}{\mathrm{d}\theta} = a_1 z + a_2 \bar{z} + b \tag{1}$$

with a_1, a_2, b general complex constants. A detailed treatment of this equation is found in current text books (and an admirable one in Sansone and Conti, 1956, II, § 1.4); here only the results will be summarized. One determines the following characteristics of (1): the discriminant $\delta = |a_1|^2 - |a_2|^2$, and the trace $\tau = 2 \operatorname{Re} a_1$. In the nondegenerate case, characterized by $\delta \neq 0$, there is a unique critical point, the solution z of $a_1 z + a_2 \bar{z} + b = 0$, viz.

$$z = \frac{-\bar{a}_1 b + a_2 \bar{b}}{\delta}\,;$$

the disposition of the solutions is then determined by the roots e of the characteristic equation

$$e^2 - \tau e + \delta = 0\,.$$

There are precisely three topologically distinct situations, the *saddle point* ($\delta < 0$), the *focus* ($\delta > 0 \neq \tau$; *stable* if $\tau < 0$, *unstable* if $\tau > 0$), and the *centre* ($\delta > 0 = \tau$). The corresponding canonic forms of (1) are, respectively,

$$\frac{\mathrm{d}z}{\mathrm{d}\theta} = \bar{z}\,, \quad \frac{\mathrm{d}z}{\mathrm{d}\theta} = z\,, \quad \frac{\mathrm{d}z}{\mathrm{d}\theta} = iz\,.$$

In the classical approach, the case of a focus is decomposed into four subcases – the true focus, the discitical node, the one-tangent node, and the two-tangent node (*loc. cit.*).

The direct generalization of equation (1) is the differential equation of algebraic type,

$$\frac{dz}{d\theta} = \sum_{n=0}^{m} \sum_{k=0}^{n} a_{kn} z^k \bar{z}^{n-k},$$

with m a positive integer and a_{kn} complex constants, and with the homogeneous algebraic form as special case,

$$\frac{dz}{d\theta} = \sum_{k=0}^{m} a_k z^k \bar{z}^{m-k}$$

(Niemyckij and Stepanov, 1949, II, § 4). A generalization of (1) with $a_2 = 0$ are the meromorphic systems

$$\frac{dz}{d\theta} = f(z)$$

with f a meromorphic function.

2.2. The general second-order autonomous differential equation is

$$\ddot{x} = f(x, \dot{x}), \tag{2}$$

with real x; this may be "reduced" to a first-order equation in R^2 by the usual device: in real coordinates x, y,

$$\frac{dx}{d\theta} = y, \quad \frac{dy}{d\theta} = f(x, y).$$

We proceed to consider some important particular cases.

If f in (2) has $f(x, y)$ independent of y, then the physical system described by (2) is called *conservative* (with one degree of freedom). A familiar instance is that of an idealized pendulum, oscillating without friction in a uniform (gravitational) field of force; the corresponding differential equation for the angular displacement φ is

$$\frac{d^2\varphi}{d\theta^2} + \frac{g}{l} \sin \varphi = 0.$$

A similar equation governs the behaviour of a non-linear LC circuit:

$$\frac{d^2\varphi}{d\theta^2} + \frac{1}{C} a(\varphi) = 0$$

where φ measures the magnetic flux, and $i = a(\varphi)$ is the current-flux characteristic.

Observe that conservative systems (with one degree of freedom) possess differential equations which are directly integrable in the elementary text-book sense.

2.3. The non-conservative systems are most interesting; they seem to originate in practice from two sources: by taking into account energy losses within the system itself, and in systems in which energy is supplied from the outside, often by some feed-back device. An instance of the former are electrical circuits containing a linear resistance. Or again, mechanical oscillating systems with friction, such as the pendulum moving freely in the atmosphere,

$$\frac{d^2\varphi}{d\theta^2} + \lambda \frac{d\varphi}{d\theta} + \frac{q}{l} \sin \varphi = 0 \tag{3}$$

(or replacing the right side by a positive constant, the equation of the starting regime of a synchronous electric motor); or a linear spring device subjected to Coulomb friction

$$\frac{d^2x}{d\theta^2} + \lambda \operatorname{sgn} \frac{dx}{d\theta} + \omega^2 x = 0 . \tag{4}$$

As an example of differential equations of systems with nonlinear feed-back, consider the two classes, that of Liénard-type equations

$$\frac{d^2x}{d\theta^2} + f(x) \frac{dx}{d\theta} + x = 0 \tag{5}$$

with the celebrated special case, the van der Pol equation

$$\frac{d^2x}{d\theta^2} - \varepsilon(1 - x^2) \frac{dx}{d\theta} + x = 0 ;$$

and Rayleigh-type equations

$$\frac{d^2x}{d\theta^2} + f\left(\frac{dx}{d\theta}\right) \frac{dx}{d\theta} + x = 0 ,$$

with the Rayleigh equation as representative case

$$\frac{d^2x}{d\theta^2} - \varepsilon\left(1 - \left(\frac{dx}{d\theta}\right)^2\right) \frac{dx}{d\theta} + x = 0 .$$

A wealth of information concerning both the general and the special cases of these equations may be found in Sansone and Conti (1956, VI), with references to the extremely extensive literature up to 1956.

2.4. In any case the equations of 2.3 may be reduced to first-order equations in R^2 by means of the general process described in 2.2. However, for Liénard-type equations another procedure is occasionally applied: in (5), set

$$F(x) = \int_0^x f(\xi)\, d\xi ,$$

and substituting

$$\frac{dx}{d\theta} + F(x) = y$$

obtain

$$\frac{dx}{d\theta} = y - F(x) , \quad \frac{dy}{d\theta} = -x .$$

Of course, the restriction to Liénard-type equations is unnecessary, and one may so treat equations of the form

$$\frac{d^2x}{d\theta^2} + f(x)\frac{dx}{d\theta} + g\left(x, \frac{dx}{d\theta}\right) = 0 ,$$

and, in particular, the so-called generalized Liénard equations

$$\frac{d^2x}{d\theta^2} + f(x)\frac{dx}{d\theta} + g(x) = 0 .$$

Observe that in the case of some equations with discontinuous coefficients, for example

$$\frac{d^2x}{d\theta^2} + \lambda \operatorname{sgn} x \frac{dx}{d\theta} + g(x) = 0 , \tag{6}$$

the described procedure still results in continuous right-hand sides of the corresponding equation in R^2,

$$\frac{dx}{d\theta} = y - |x| , \quad \frac{dy}{d\theta} = -g(x) .$$

3. GENERALIZATION: NON-EUCLIDEAN CARRIER SPACES

In this section, the term "carrier" is used rather informally to describe the set within which the differential equation functions; for example, for (1) below, the carrier is the set G. For autonomous equations, *phase-space* is often used synonymously.

3.1. The autonomous differential equations described in Section 1 were of the form

$$\frac{dx}{d\theta} = f(x) \tag{1}$$

with continuous $f : G \to \mathsf{R}^n$, and G an open subset of R^n. In some cases it is entirely natural and quite useful to study, along with (1), an allied object on a carrier set with a different coordinate structure.

Thus, consider an equation such as (3) in 2.3. It may be noticed that, together with any solution φ, there are also solutions $\varphi + 2\pi$, $\varphi + 4\pi$, etc. Thus, vaguely speaking, the configuration of solutions is reproduced over and over again; and to obtain complete information on all solutions, one need only study a rather restricted subset of the solutions.

3.2. More generally, take an autonomous equation in R^2 written in real coordinates x, y as

$$\frac{dx}{d\theta} = f(x, y), \quad \frac{dy}{d\theta} = g(x, y), \tag{2}$$

and assume that both f, g have period 2π in y, i.e.

$$f(x, y + 2\pi) = f(x, y), \quad g(x, y + 2\pi) = g(x, y)$$

for all $(x, y) \in \mathsf{R}^2$. Now set $z = e^{iy}$ and consider

$$\frac{dx}{d\theta} = f_0(x, z), \quad \frac{dz}{d\theta} = iz\, g_0(x, z) \tag{3}$$

where f_0 and g_0 are defined by

$$f_0(x, e^{iy}) = f(x, y), \quad g_0(x, e^{iy}) = g(x, y).$$

It is easily verified that, for each solution (x, y) of (2), (x, e^{iy}) is a solution of (3). Conversely, let (x, z) be a solution of (3) in the obvious sense that, for some open interval $I \subset \mathsf{R}^1$,

$$x : I \to \mathsf{R}^1, \quad z : I \to \mathsf{R}^2, \quad |z(\theta)| \equiv 1,$$

$$\frac{d}{d\theta} x(\theta) = f_0(x(\theta), z(\theta)), \quad \frac{d}{d\theta} z(\theta) = i\, z(\theta)\, g_0(x(\theta), z(\theta))$$

for all $\theta \in I$. Then one has $z(\theta) = e^{iy(\theta)}$ for some continuous $y : I \to \mathsf{R}^1$, and it is easily verified that the resulting (x, y) is a solutions of (2). This establishes a correspondence between solutions of (2) in Euclidean 2-space R^2, and solutions of (3) on the 2-cylinder $\mathsf{R}^1 \times \mathsf{S}^1$; obviously solutions of (2) which differ by a period $2k\pi$ in the second coordinate go over into a single solution of (3).

3.3. As an example, the mentioned second-order equation of 2.3,

$$\frac{d^2\varphi}{d\theta^2} + \lambda \frac{d\varphi}{d\theta} + \frac{q}{l} \sin \varphi = 0$$

is first reduced to a first-order equation in R^2 ($y = \varphi$, $x = d\varphi/d\theta$),

$$\frac{dx}{d\theta} = -\lambda x - \frac{g}{l} \sin y\,, \quad \frac{dy}{d\theta} = x\,;$$

and using $z = e^{iy}$ etc., as described in 3.2, there results

$$\frac{dx}{d\theta} = -\lambda x - \frac{g}{2li}\left(z - \frac{1}{z}\right), \quad \frac{dz}{d\theta} = izx\,.$$

Formally, this would also define an equation in $R^1 \times (R^2 - (0))$, with the cylinder $R^1 \times S^1$ as integral manifold. More generally, (3) admits as integral manifold any cylinder

$$\{(x, z) : x \in R^1, |z| = \varrho\}$$

with $\varrho > 0$; this follows from

$$\frac{d}{d\theta} |z(\theta)|^2 = \frac{dz}{d\theta} \bar{z} + z \overline{\frac{dz}{d\theta}} = i|z|^2 g_0 - i|z|^2 g_0 \equiv 0\,.$$

3.4. An analogous procedure may be used in some other situations. Thus if in (2) the f, g are periodic in both variables, then both variables may be independently subjected to the suggested procedure; there results a differential equation on the 2-torus $S^1 \times S^1$ (for a detailed treatment see Coddington and Levinson, 1955, XVII).

3.5. As an example of another type; consider a homogeneous algebraic (autonomous differential, cf. 2.1) equation in R^2

$$\frac{dz}{d\theta} = \sum_{k=0}^{n} a_k z^k \bar{z}^{n-k}\,;$$

now make the carrier R^2 compact by adjoining a point at infinity, and attempt to extend the differential equation over this larger carrier space. In any case, the behaviour of large solutions of (4) is better treated by first transforming $z = 1/w$; after rearrangement one obtains

$$\frac{dw}{d\theta} = -\frac{1}{|w|^{2n}} \sum_{k=0}^{n} a_k w^{n-k+2} \bar{w}^k\,.$$

In particular, from $dz/d\theta = az^n$ there results $dw/d\theta = -aw^{2-n}$, and for $n = 0, 1, 2$ this is again homogeneous algebraic. In the general case, the real-valued factor $|w|^{-2n}$

may be eliminated by re-parametrizing the solutions (cf. VI, 1.13); thus if one is only interested in "geometrical" properties of the solution, it suffices to consider

$$\frac{\mathrm{d}w}{\mathrm{d}\theta} = -\sum_{k=0}^{n} a_k w^{n-k+2}\overline{w}^k ,$$

which is again algebraic homogeneous.

Another example, due to Poincaré, is described in detail in Lefschetz (1951, IX, §5); here the original carrier R^2 is compactified by forming the associated projective plane.

3.6. The examples described in the present section are lacking in one respect: it remains to be specified what is meant by a differential equation on the carrier spaces introduced, and by solutions of such equations. The careful reader must have noticed that, without precise definitions, even the elementary results of 3.2 and 3.5 remain in the air. However, it is not difficult to exhibit these definitions.

Let P be a differential n-manifold; a differential equation in P is defined by a continuous map $f : D \to \mathsf{R}^n$ with D an open subset of $P \times \mathsf{R}^1$, and written as

$$\frac{\mathrm{d}x}{\mathrm{d}\theta} = f(x, \theta) . \tag{5}$$

In the case that $D = P \times \mathsf{R}^1$ and f is independent of θ, the equation is written as

$$\frac{\mathrm{d}x}{\mathrm{d}\theta} = f(x) ,$$

and f is said to define an autonomous differential equation on P.

Next, let $x : I \to P$ be a continuous map with $I \subset \mathsf{R}^1$ an open interval, and let C be any admissible differential coordinate system on P. Then x is called a solution of (5) if, for any coordinate $c \in C$,

$$\frac{\mathrm{d}}{\mathrm{d}\theta}\, c^{-1}(x(\theta)) = f(x(\theta), \theta) \tag{6}$$

whenever $c^{-1}(x(\theta))$ is defined. In the autonomous case this may be formulated more vividly as requiring $c^{-1}x$ to be a solution of fc (in the sense of 1.1), for all $c \in C$.

Observe that (6) implies that if $x(\theta)$ is in the coordinate neighbourhoods of two coordinate functions $c_1, c_2 \in C$, then

$$c_1^{-1}\, x(\theta) = (c_1^{-1} c_2)\,(c_2^{-1}\, x(\theta)) ,$$

and on differentiating,

$$f(x(\theta), \theta) = \frac{\partial c_1^{-1}\, c_2}{\partial x}\, f(x(\theta), \theta) .$$

3.7. Possibly it may be useful to observe the functioning of this definition on the example described in 3.2. The differential 2-manifold concerned is the 2-cylinder $\mathsf{R}^1 \times \mathsf{S}^1$; a simple coordinate system consists of two maps only, c_1 and c_2, defined by

$$c_1(x, y) = (x, e^{iy}) \quad \text{for} \quad x \in \mathsf{R}^1, \quad -\pi < y < \pi,$$
$$c_2(x, y) = (x, e^{iy}) \quad \text{for} \quad x \in \mathsf{R}^1, \quad 0 < y < 2\pi.$$

Note that $c_i^{-1} c_j$ is the identity map (in each component of the corresponding domain). Now consider the autonomous differential equation on $\mathsf{R}^1 \times \mathsf{S}^1$

$$\frac{\mathrm{d}}{\mathrm{d}\theta}(x, z) = (f_0(x, z), g_0(x, z)), \tag{7}$$

where f_0, g_0 are the maps defined in 3.2; we observe that the second component is $g_0(x, z)$ and not the $iz\, g_0(x, z)$ appearing in (3). According to 3.6, a continuous map $(x, z) : I \to \mathsf{R}^1 \times \mathsf{S}^1$ is a solution of (7) iff the following holds: assuming that, say, $(x(\theta), z(\theta)) \in \textit{image}\ c_1$ for an $\theta \in I$, one has

$$c_1^{-1}(x(\theta), z(\theta)) = (x(\theta), y(\theta)) \quad \text{with} \quad y(\theta) = -i \log z(\theta),$$

and then (6) reduces to

$$\frac{\mathrm{d}}{\mathrm{d}\theta} x(\theta) = f_0(x(\theta), z(\theta)) = f(x(\theta), y(\theta)),$$

$$\frac{\mathrm{d}}{\mathrm{d}\theta} y(\theta) = g_0(x(\theta), z(\theta)) = g(x(\theta), y(\theta)),$$

as expected.

3.8. Having described the generalization of differential equations to carriers which are merely locally Euclidean, a cursory glance at further reasonable types of carrier space may be in place.

Let there be given real constants $\alpha, \alpha_1, \alpha_2, \ldots, \alpha_n$ and a map $y_0 : \mathsf{R}^n \to \mathsf{R}^1$; write $a = [\alpha_k] \in \mathsf{R}^n$ and $\nabla = [\partial/\partial\xi_1, \ldots, \partial/\partial\xi_n]$, the nabla operator in R^n; and consider the partial differential initial value problem

$$\frac{\partial y}{\partial \theta} = \alpha y + (a, \nabla y) = \alpha y + \sum_1^n \alpha_k \frac{\partial y}{\partial \xi_k}, \tag{8}$$
$$y(x, 0) = y_0(x).$$

Obviously, if y_0 is of class C, the unique solution is given by

$$y(x, \theta) = e^{\alpha\theta}\, y_0(x + a\theta);$$

for fixed $\theta \in \mathsf{R}^1$, this again represents a map of class C. Thus it seems natural to term (8) a differential equation in the set of C maps $\mathsf{R}^n \to \mathsf{R}^1$; note that (8) also exhibits an autonomness property in that, if y is a solution of (8) and $\theta_0 \in \mathsf{R}^1$, then so is the map z, defined by

$$z(x, \theta) = y(x, \theta + \theta_0) = e^{\alpha\theta_0}\, y(x, \theta)$$

(cf. 1.14). For further information concerning this and allied examples see Zubov (1957, I, §5 and V). An admirable survey of linear differential equations in Banach spaces appears in Kreĭn (1964, III).

3.9. The final example concerns differential equations with retarded argument (or time lag). Let there be given a continuous $f : \mathsf{R}^n \times \mathsf{R}^n \to \mathsf{R}^n$, a positive $\tau \in \mathsf{R}^1$, and a continuous $x_0 : [0, \tau) \to \mathsf{R}^n$; and consider the problem of determining a map $x : \mathsf{R}^1 \to \mathsf{R}^n$ so as to satisfy

$$\frac{\mathrm{d}}{\mathrm{d}\theta}\, x(\theta) = f(x(\theta), x(\theta - \tau)), \qquad (9)$$

$$x \mid [0, \tau) = x_0 .$$

Under rather weak conditions on f and x_0 (related to the unicity and extendability conditions in Section 1), the problem (9) does have a unique solution; it even exhibits an autonomness property in that, if x satisfies the differential relation in (9), then so does y defined by $y(\theta) = x(\theta + \theta_0)$ for any fixed $\theta_0 \in \mathsf{R}^1$.

Hence (9) may also be interpreted as operating on the set of continuous maps $[0, \tau) \to \mathsf{R}^n$, in the following manner. Take any continuous $x_0 : [0, \tau) \to \mathsf{R}^n$ and any $\theta_0 \in \mathsf{R}^1$, and determine the solution x of (9); then shift x along the θ-axis by $-\theta_0$, and partialize to $[0, \tau)$, obtaining a map $y(x_0, \theta_0) : [0, \tau) \to \mathsf{R}^n$ again. For definiteness, the map $y(x_0, \theta_0)$ is defined by

$$y(x_0, \theta_0)\,(\theta) = x(\theta + \theta_0) \quad \text{for} \quad 0 \leqq \theta < \tau .$$

A survey of results on differential equations with retarded argument appears in El'sgolc (1955, V).

4. GENERALIZATION: NON-DIFFERENTIABILITY

Evidently, non-differentiability of solutions of

$$\frac{\mathrm{d}x}{\mathrm{d}\theta} = f(x, \theta)$$

is closely connected with discontinuity of the map f; one may then say that the differential equation is discontinuous.

4.1. Such equations have already appeared among the examples given previously: (4) and (6) in Section 2. The physical background of such phenomena in these and similar examples is that the physical laws which govern the behaviour of the system described by the equation are inherently discontinuous (as in the case of Coulomb friction).

In other situations, the underlying physical laws do have the appropriate smoothness properties, but conditions subsequently imposed on the system lead to non-differentiability of the motions. To illustrate this, consider a simple harmonic oscillator, described by

$$\ddot{\varphi} + \omega^2 \varphi = 0$$

subjected to the constraint $|\varphi| \leqq \alpha$ (ω, α positive constants). The resulting system may then be described by

$$\frac{d\varphi}{d\theta} = \begin{cases} \psi & \text{for } |\varphi| < \alpha, \\ -\psi & \text{for } |\varphi| \geqq \alpha, \end{cases} \quad \frac{d\psi}{d\theta} = -\omega^2 \varphi ;$$

and motions with $\varphi(0) = 0$ and large $\dot{\varphi}(0)$ will then have a discontinuous derivative.

A third instance of discontinuous equations occurs on supplying an on-off feedback to a system with otherwise smooth behaviour; an example of this type will be examined in the following section.

4.2. An early source of discontinuous differential equations in the physical sciences was the practice of studying the transient behaviour of, say, linear systems by noting the reaction of the system to an input η of the form

$$\eta(\theta) = \begin{cases} 0 & \text{for } \theta < 0, \\ 1 & \text{for } \theta \geqq 0. \end{cases}$$

In the case of a second-order lumped-parameter system, the differential equation of the corresponding output is

$$L\ddot{y} + R\dot{y} + \frac{1}{C} y = \eta ,$$

i.e. discontinuous.

4.3. It is, of course, necessary to re-define the concept of solution so as to cover the case of discontinuous equations.

Thus, assume given a map $f : D \to \mathsf{R}^n$, $D \subset \mathsf{R}^n \times \mathsf{R}^1$; again, f will be said to define the differential equation

$$\frac{dx}{d\theta} = f(x, \theta) ; \tag{1}$$

for continuous f and open D this only repeats 1.1. A continuous map $x : I \to \mathsf{R}^n$

with $I \subset \mathsf{R}^1$ an open interval is called a *generalized solution* of (1) if, for any θ and θ_0 in I,

$$x(\theta) = x(\theta_0) + \int_{\theta_0}^{\theta} f(x(\lambda), \lambda)\, \mathrm{d}\lambda \tag{2}$$

with the Lebesgue integral on the right-hand side.

Observe that if f is continuous at $(x(\theta'), \theta')$, or at least $f(x(\theta), \theta)$ is continuous at an $\theta' \in I$, then (2) implies

$$\frac{\mathrm{d}}{\mathrm{d}\theta} x(\theta) = f(x(\theta), \theta) \quad \text{at} \quad \theta = \theta' .$$

Therefore the present definition is indeed a generalization of that in 1.1 for continuous maps f.

In the general case, with f arbitrary, there is no existence theorem which is parallel to the Cauchy-Peano theorem 1.3; this may be seen in the following section. However, if the discontinuity of f is due largely to its second variable, θ, one has the celebrated result (Kamke, 1951, A, §2).

4.4. *Carathéodory theorem. Let* $f : \mathsf{R}^n \times I \to \mathsf{R}^n$ ($I \subset \mathsf{R}^1$ *an open interval*) *be a map with the following properties.*

1. *For fixed* $x \in \mathsf{R}^n$, $f(x, \theta)$ *is Lebesgue measurable in* θ; *there exists a subset* $A \subset I$ *with mes* $A =$ *mes* I *such that*

2. *for fixed* $\theta \in A$, $f(x, \theta)$ *is continuous in* x;

3. *to integral* $m > 0$ *there exist Lebesgue summable functions* $\tau_m : \mathsf{R}^n \to \mathsf{R}^1$ *with*

$$\|f(x, \theta)\| \leqq \tau_m(x) \quad \text{for} \quad x \in \mathsf{R}^n, \quad \|x\| \leqq m, \quad \textit{and} \quad \theta \in A .$$

Then, to any initial data $(x_0, \theta_0) \in \mathsf{R}^n \times I$, *there exists an absolutely continuous map* $x : I \to \mathsf{R}^n$ *such that* $x(\theta_0) = x_0$ *and that* x *is a generalized solution of* (1).

4.5. *Remarks.* Under analogous assumptions, reasonable conditions ensuring unicity and continuous dependence on initial data and parameters may also be given (cf. Coddington and Levinson, 1955, II, 4.2 for $n = 1$). Further generalizations in this direction of the differential equation concept have been given by Kurzweil and his associates, with special emphasis on results connected with dependence on parameters in the equation. A survey appears in Kurzweil, 1962.

In the autonomous case, the Carathéodory conditions (and also Kurzweil's) reduce to ordinary continuity; hence they do not cover the situation of a discontinuous autonomous equation. Since, in the present treatise, we shall mostly be concerned with autonomous equations and their abstract counterparts, this direction of generalizing will not be pursued further.

5. DISCONTINUOUS AUTOMATIC CONTROL

This section is devoted to the introductory study of a simple class of linear second order regulating systems, supplied with a non-linear discontinuous feed-back of various types. The exposition is based on a reformulation of Flügge-Lotz (1953, I–V and VI); for further results and references, including applications, see Neymark (1961).

5.1. The starting-point is the linear equation with constant coefficients

$$\ddot{\psi} + 2\alpha\dot{\psi} + \psi = 0 \tag{1}$$

with

$$0 < \alpha < 1 .$$

This will be termed the *unregulated equation.* Setting $\beta = \sqrt{(1 - \alpha^2)}$ one has, as solutions of (1)

$$\psi(\theta) = ce^{p\theta} + \bar{c}e^{\bar{p}\theta} \tag{2}$$

with $p = -\alpha + i\beta$, and c an arbitrary complex constant; observe that

$$|p| = 1 , \quad \operatorname{Re} p = -\alpha .$$

A slightly unconventional passage from (1) to a first-order equation in R^2 is effected using the transformation

$$z = \dot{\psi} - \bar{p}\psi . \tag{3}$$

yielding

$$\dot{z} = \ddot{\psi} - p\dot{\psi} = -(2\alpha + \bar{p})\dot{\psi} - \psi = p\dot{\psi} - p\bar{p}\psi = p(\dot{\psi} - \bar{p}\psi) ,$$

$$\frac{dz}{d\theta} = pz . \tag{4}$$

The trajectories of (4) are logarithmic spirals with focus in the origin; the solutions (2) describe damped oscillations with frequency β and damping coefficient α. In the z-plane, the ψ, $\dot{\psi}$ appearing in (3) may be treated as oblique coordinates; the positive $\dot{\psi}$-axis coincides with the semiaxis of positive reals.

5.2. Next, assume that onto the physical system corresponding to (1) there is superimposed a two-position feedback device, whose effect may be described by the following modification of (1):

$$\ddot{\psi} + 2\alpha\dot{\psi} + \psi = \operatorname{sgn} \varphi ; \tag{5}$$

here the *regulator* φ is to depend, essentially, on the immediate state of the regulated system, i.e. on the values of ψ and $\dot{\psi}$; the corresponding formula is called the *regulator equation.*

On performing the transformation (3) as above, there results, in place of (4),

$$\frac{dz}{d\theta} = pz + \operatorname{sgn}\varphi = p(z + \bar{p}\operatorname{sgn}\varphi), \tag{6}$$

with formal "solutions"

$$z(\theta) = ce^{p\theta} - \bar{p}\operatorname{sgn}\varphi. \tag{7}$$

One may at least conclude that, as long as φ is non-zero and preserves sign, trajectorial arcs are arcs of logarithmic spirals with foci at $\pm\,\bar{p}$; these are indicated in Fig. 2.

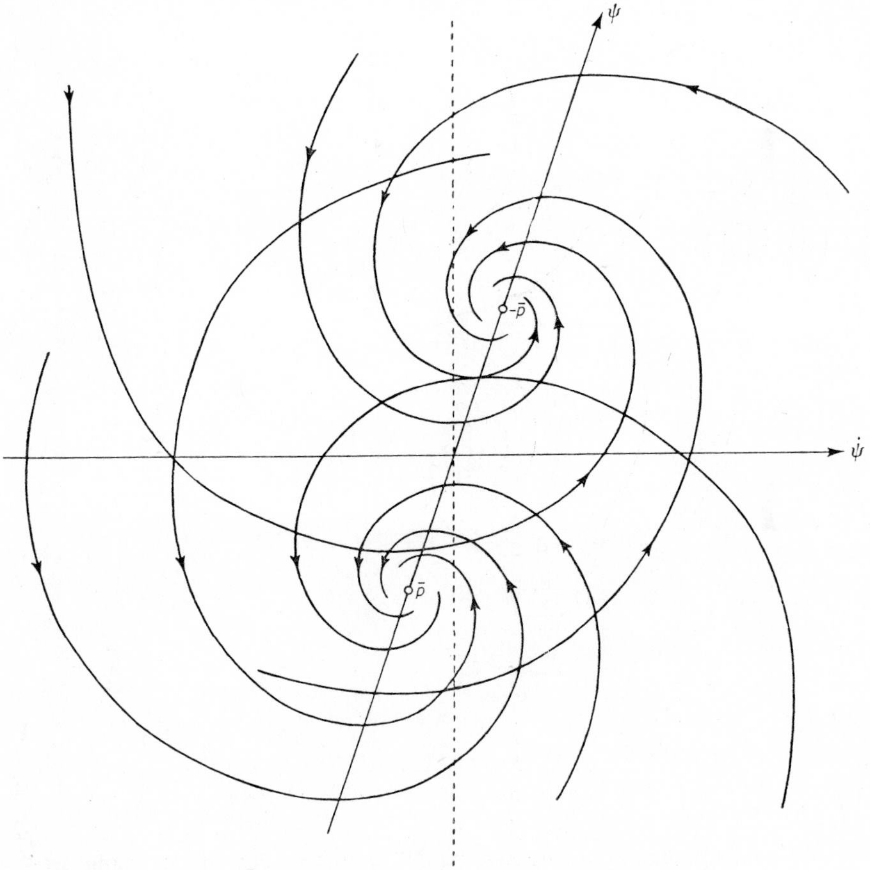

Fig. 2. Permissible form of trajectories.

5.3. The actual behaviour of the system described by (5) will, of course, depend on the regulator equation. Assume, as a first example, that

$$\varphi = \psi + \varkappa\dot{\psi}$$

with $\varkappa$ some positive real constant. In the terminology of Flügge-Lotz (1953) this is case B of a two-position regulating regime. The geometry of the trajectories is sketched in Fig. 3, where one may note the role played by the switch-curve with equation $\varphi = 0$, i.e. straight line with direction $1 + \varkappa\bar{p}$,

$$z = (1 + \varkappa\bar{p})\,\lambda\,, \quad \lambda \in \mathrm{R}^1\,. \tag{8}$$

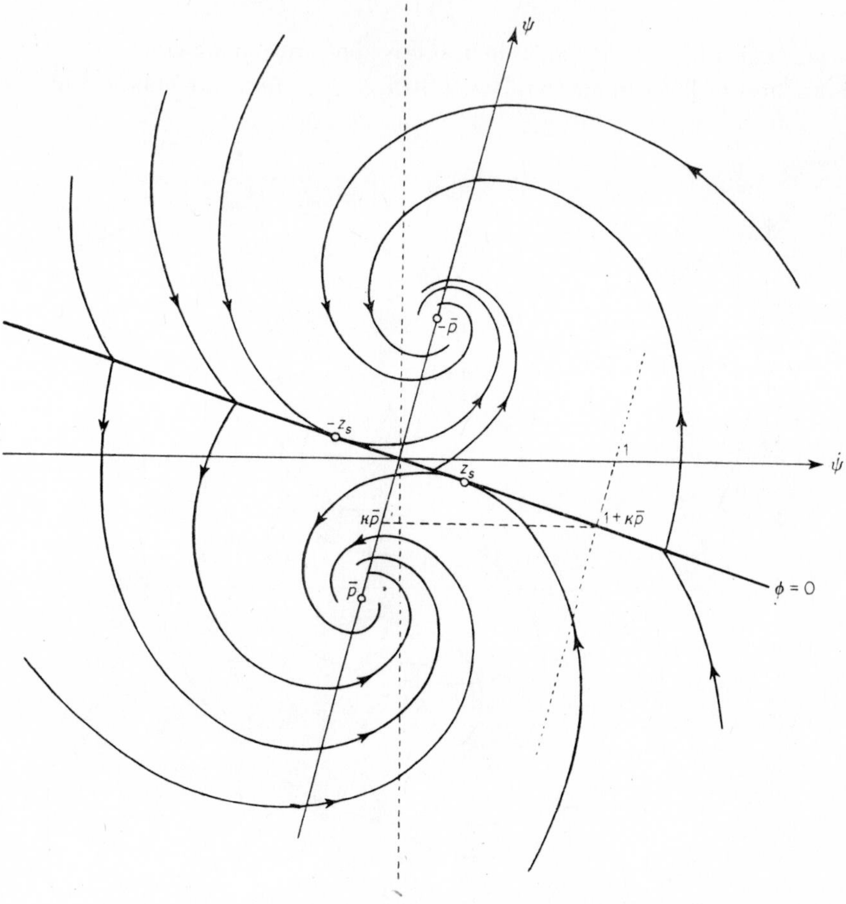

Fig. 3. Trajectories of a self-regulating system.

5.4. Two phenomena are of interest here. The first is that, in traversing the switch curve, the continuous trajectories suffer a discontinuity of their tangents; hence the corresponding solutions are non-differentiable. Evidently the situation is well-covered by the definition of generalized solution introduced in 4.3, and should not lead to any conceptual difficulties.

The second is the rather confusing behaviour on the switch curve between the points marked z_S and $-z_S$. These may be defined as the points at which trajectories are tangent to the switch-curve. Hence, at $z = z_S$, $dz/d\theta$ is parallel to $1 + \varkappa\bar{p}$; i.e. on setting $\delta = \operatorname{sign} \varphi = \pm 1$,

$$0 = \begin{vmatrix} pz + \delta, & \bar{p}\bar{z} + \delta \\ 1 + \varkappa\bar{p}, & 1 + \varkappa p \end{vmatrix} = zp(1 + \varkappa p) - \bar{z}\bar{p}(1 + \varkappa\bar{p}) + \delta\varkappa(p - \bar{p}) .$$

Since $z = z_S$ is also on the line (8),

$$0 = \begin{vmatrix} z, & \bar{z} \\ 1 + \varkappa\bar{p}, & 1 + \varkappa p \end{vmatrix} = z(1 + \varkappa p) - \bar{z}(1 + \varkappa\bar{p}) .$$

From these two equations there follows

$$z_S = \frac{-\delta\varkappa}{1 + \varkappa p} .$$

If $\varkappa$ is allowed to vary over the reals, the corresponding points z_S map out two symmetric circles, tangential to the axis of reals at the origin (for $\varkappa = 0$), passing through the points

$$\frac{-\delta}{p} = -\delta\bar{p} = \pm\bar{p}$$

(for $\varkappa \to +\infty$), with common radii $1/2\beta$ and centres $\pm i/2\beta$.

5.5. Now consider the portion S of the switch-curve between the points $\pm z_S$. The situation evident from Fig. 3 may be described, in rather vague terms, by saying that each non-end-point of S is the initial point of two trajectories, and these move away from S in roughly opposite directions. In particular, no generalized solution has initial data (z_0, θ_0) with z_0 a non-end-point of S, since each solution $x : I \to \mathsf{R}^2$ must have open domain I.

Three ways of avoiding this difficulty suggest themselves. The first, and simplest, is to study the equation (6) only in the open set $\mathsf{R}^2 - S$. However, this may not be satisfactory. A compromise is to omit only the end-points $\pm z_S$ of S and duplicate the remaining portion of S, assigning one specimen to the lower and the second to the upper half-planes. The resulting object is a 2-manifold with boundary, and (6) may be considered on this carrier set. The boundary points are so-called *s-points*, i.e. start-points of some semi-trajectory (for a precise definition, see II, 2.4). This configuration is treated in VI,4.

A more sophisticated approach would be to take several specimens not only of S but of the whole plane, define equations related to (6) on each specimen, and then identify some portions of these in a suitable manner; intuitively, the Riemann surface of $\sqrt{(z^2 - 1)}$ might be a useful carrier to employ. Chapter V is devoted to this type of construction in a rather more abstract setting; a surprising result is that, in a certain sense, the extended carrier space is uniquely determined (cf. V,2.5).

5.6. Naturally, one inquires about the presence of cycles of (6). Observe that both the unregulated equation (4) and the regulated equation (6) are centrally symmetric, and possibly it is obvious that each cycle of (6), if such exists, must encircle the origin. From the Poincaré symmetry principle it follows that there exists a cycle iff some trajectory intersects the switch-curve at points z_c and $-z_c$, whereupon $\pm z_c$ are on a cycle. Since it may be assumed that the characteristics are consecutive, (7) may be applied; writing $\delta = \operatorname{sgn} \varphi = \pm 1$ again, one has

$$z_c = \left[ce^{p\theta} - \bar{p}\delta\right]_{\theta=0}, \quad -z_c = \left[ce^{p\theta} - \bar{p}\delta\right]_{\theta=\theta},$$

and on eliminating c,

$$z_c = \bar{p}\delta \frac{1 - e^{p\theta}}{1 + e^{p\theta}}. \tag{9}$$

This is the parametric equation (parameter: θ) of a centrally symmetric curve, spiralling around $\mp \bar{p}$; in particular, it must intersect the switch-curve (at least for some values of K) since it separates $\bar{p}$ from $-\bar{p}$. Finally then, there exist cycles of (6), whose intercepts with the switch-curve coincide with those of (9), and values of $\theta > 0$ corresponding to such intersections are semi-periods.

Evidently, several of these steps would not stand adverse criticism. In particular, the Poincaré symmetry principle was enounced for a different type of symmetry, and only for continuous equations in R^2. However, it is hoped that the reader will at least admit the plausibility of the argument. All these defects will be remedied in subsequent chapters.

5.7. Finally, we present a brief survey of other types of on-off feed-back, but only from the point of view of the questions treated in the present chapter. The physical motivation is the attempt at a more accurate description of realistic feed-back devices.

If the regulator equation has the form

$$\varphi = \begin{cases} \psi + \varkappa\dot{\psi} & \text{if } |\psi + \varkappa\dot{\psi}| > \gamma \\ 0 & \text{if } |\psi + \varkappa\dot{\psi}| \leqq \gamma \end{cases}$$

($\varkappa$, γ positive constants), the regulator is said to exhibit a *dead zone.* The treatment is similar to that preceding except the switch curve now consists of two parallel lines $\psi + \varkappa\dot{\psi} = \pm\gamma$.

Further physically interesting cases arise on assuming that the regulator depends not only on the immediate state of the system, but also on some of its previous history. This situation arises if the regulator passes from one of its two states to the other (i.e. "switches") a fixed positive time τ after $\psi + \varkappa\dot{\psi}$ had assumed zero value, a typical case of *time-lag*. The switch curve equation is easily obtained as follows: setting $\delta = \operatorname{sgn} \varphi$, $z = ce^{p\tau} - \bar{p}\delta$, $z_0 = c - \bar{p}\delta$, one has

$$z_0(1 + \varkappa p) - \bar{z}_0(1 + \varkappa\bar{p}) = 0,$$

and the required equation is

$$ze^{-p\tau}(1 + \varkappa p) - \bar{z}e^{-\bar{p}\tau}(1 + \varkappa\bar{p}) = \delta(\varkappa + \bar{p})(1 - e^{-p\tau}) - \\ - \delta(\varkappa + p)(1 - e^{-\bar{p}\tau}).$$

In another situation, switching occurs if $|\psi + \varkappa\dot{\psi}|$ first reaches a given value γ having first passed through zero; this is the phenomenon called *hysteresis* in Flügge-Lotz (1953).

As in the first case, in each of these one again has the problem of choosing an appropriate carrier space for the equation, for example, an open subset of R^2 or a suitable Riemann-surface-like covering space. Thus, it appears both useful and interesting to study discontinuous differential equations on 2-dimensional (but non-planar) carriers.

6. GENERALIZATION: DYNAMICAL SYSTEMS AND FLOWS

6.1. The concept of a dynamical system appeared, implicitly, in papers on topological and metric properties of differential equations in the period marked by the names of H. Poincaré and G. D. Birkhoff. The explicit definition was given in 1931 by A. A. Markov (see references to chapter V in Niemyckij and Stepanov, 1949), and essentially is reproduced below. Like many of the fundamental definitions of that period, the concept suffers from the over-emphasis of metric spaces (among topological spaces) which was then current; it may be observed that the tendency is still diligently perpetuated.

A *dynamical system* in a metric space P is a continuous map $f : P \times \mathsf{R}^1 \to P$ with the *initial-value property*

$$f(x, 0) = x,$$

and the *group property*

$$f(f(x, \theta), \theta') = f(x, \theta + \theta')$$

for all $x \in P$ and $\theta, \theta' \in \mathsf{R}^1$ (Niemyckij and Stepanov, 1949). (In subsequent chapters, the term dynamical system is used in a slightly different sense, and the present concept would be termed a continuous global dynamical system on the metrizable space P.)

The connection with differential equation theory is established by noting that, in the situation of 1.16,

$$f(x, \theta) = {}_\theta T_0 x$$

defines a dynamical system f on G; see 1.17 and 1.9 for continuity of f, with G considered a metric subspace of $\mathsf{R}^n \supset G$. Thus dynamical systems are a generalization, to metric carrier spaces, of autonomous differential equations in R^n with unicity and extendability. Observe that differential n-manifolds as carriers may be admitted, and it can be shown that the same holds for discontinuous equations (again, autonomous with unicity, extendability, but local existence of solutions must be assumed). The systems on function spaces described in 3.8–9 also come under this description.

6.2. Naturally there arises the question as to whether some of these restictions (metric carrier, extendability, unicity, autonomness) can be lifted in some reasonable manner. Let us consider these in turn.

From the definition itself, it is obvious that a formally identical formulation is meaningful even if P is merely a topological space; and it seems probable that most of the now classical dynamical system theory, including the Lyapunov method, can be carried over at least to uniform spaces.

As concerns omission of the extendability hypothesis, this is one of the focal points of interest in the present book.

6.3. A rather important question is that of obtaining an object, of the same degree of generality as dynamical systems, which would be a generalization of non-autonomous differential equations. A reasonable candidate is the following concept.

A *flow* in a topological space P is a continuous map $f : P \times \mathsf{R}^1 \times \mathsf{R}^1 \to P$ with the *initial-value property*

$$f(x, \theta, \theta) = x ,$$

and the *composition property*

$$f(f(x, \theta, \theta''), \theta'', \theta') = f(x, \theta, \theta')$$

for all $x \in P$ and $\theta, \theta', \theta'' \in \mathsf{R}^1$.

With the assumptions and notation of 1.13, it may be easily verified that

$$f(x, \beta, \alpha) = {}_\alpha T_\beta x$$

defines a flow f in G. Hence flows generalize differential equations with unicity and extendability. As concerns the connection with dynamical systems, if $g : P \times \mathsf{R}^1 \to \to P$ is a dynamical system, then

$$f(x, \theta, \theta') = g(x, \theta' - \theta)$$

defines a flow f; obviously the latter is *stationary* (this term seems to be preferred to *autonomous*) in the sense that

$$f(x, \theta, \theta') = f(x, \theta + \lambda, \theta' + \lambda)$$

for all $x \in P$ and $\theta, \theta', \lambda \in \mathsf{R}^1$.

6.4. In attempting to eliminate the unicity condition in 6.3, several definitions have been suggested (cf. Zubov, 1957, IV; Seibert, 1963). The basic idea in these is to replace, in the definition of a flow, the *point* $f(x, \theta, \theta')$ by a subset of P; intuitively, this latter is to consist of all points $y(\theta')$ on all the solutions y which satisfy $y(\theta) = x$. If then each $f(x, \theta, \theta')$ reduces to a singleton, one has the concept of a flow as above; if not, the resulting object may be called a *generalized flow* (*loc. cit.*). Apart from suggesting that, possibly, generalized flows in P might be treated as ordinary flows of a special type in the set of all subsets of P (or in a suitable subset), this subject will not be pursued further in the present book.

6.5. Generalizations in other directions may also be indicated. Thus, in the definition of flows one may require that f be defined only on $P \times \mathsf{R}^+ \times \mathsf{R}^+$, i.e. that the evolution of the system in "time" θ is specified only for $\theta \geqq 0$, (Zubov, 1957). Similarly for dynamical systems, cf. 5.5 and the phenomenon of s-points; this situation will also be treated in subsequent chapters, under the term *semi-dynamical systems.*

Another possibility is suggested by Markov chains, and also by some results on Poisson stability (cf. Niemyckij and Stepanov, 1949, V, §4, 16) and on autonomous differential systems on the 2-torus (Coddington and Levinson, 1955); in the definition of flows and dynamical systems, allow the time θ to vary merely over the integers (or only the non-negative integers). One then has, in particular, that the system of translations

$$\{{}_nT_m \mid n, m \in \mathsf{C}^1, {}_nT_m x = f(x, m, n),$$

is completely determined by the simple transitions

$$\{{}_nT_{n-1} \mid n \in \mathsf{C}^1\}$$

since, for $n > m$,

$${}_nT_m = {}_nT_{n-1}\, {}_{n-1}T_{n-2} \cdots {}_{m+1}T_m \,.$$

Except for a few results in Chapter VII, these discrete flows will not be treated further here. An allied generalization exists when "time" θ is allowed to vary over an arbitrary topological group G instead of R^1. For dynamical systems of this type (often termed *transformation groups*) many of the classical dynamical concepts have been introduced and studied (Gottschalk and Hedlund, 1955).

6.6. In this and the preceding sections, a summary description has been given of some developments in autonomous differential equation theory; these may be indicated in the following diagram:

		differential manifolds		
continuous equations in R^n	↗	as carriers	↘	dynamical systems on topological spaces
	↘	generalized solutions	↗	

It is suggested that the last step is possibly over-large, having passed over an intermediate stage where many interesting and quite specific results may be had: dynamical systems on topological n-manifolds, and in particular dynamical systems on 2-manifolds.

A concise description of the objects studied in the remaining portions of this book may now be given: dynamical systems are suitably generalized to cover the case of non-extendable solutions, considering the carrier space first as an abstract set, and then considering a topological space with progressively stricter conditions on its structure.

CHAPTER II

Abstract Dynamical Systems

This chapter introduces and develops the fundamentals of the theory of abstract dynamical systems. The adjective "abstract" is not only intended to contrast with "concrete" or differential dynamical systems, but also to emphasize the absence of any other structure, whether differential, topological or uniform, on the carrier set of the dynamical system.

In comparison with the following chapters, the contents of the present one are rather elementary: the first deeper theorems of abstract dynamical system theory are deferred until Chapter III and Section 1 of Chapter V.

Parallel to local dynamical systems, there are introduced local semi-dynamical systems. The latter no doubt deserve more intense study than that meted out here, where they are treated only as an intermediate and auxiliary concept. Indeed, results are presented on general semi-dynamical systems only if they can be proved with no further effort; in the remaining cases, only semi-dynamical systems with -unicity (cf. 2.4) are considered: these are later shown to be intimately connected with dynamical systems proper (Chapter V).

Sections 1 and 2 consist essentially of the definitions of ld systems and lsd systems; the connection between these concepts is established in theorem 2.9. Section 3 contains several fundamental constructions connected with dynamical systems, including direct products and factorization. Section 4 introduces the notions of trajectories, cycles and critical points familiar from differential equation theory. In Section 5 there is an exposition of elementary properties of invariant sets and related notions; these may be considered as the counterpart of integral manifolds. The concluding Section 6 introduces the sections, an auxiliary concept also studied in Chapter VI and extensively applied from Chapter VII onward.

Category-theoretical terms may be noticed, starting in Section 3. It should be emphasized that category theory is used throughout this book only as a source of unified terminology and concepts, and not as a proof method; thus the reader not familiar with categories and functors is requested not to be discouraged in any way by their appearance.

1. LOCAL DYNAMICAL SYSTEMS

A local dynamical system on a set P is a mapping T (of a special type described axiomatically in 1.1 below) taking a subset of $P \times \mathsf{R}^1$ into P. The value which T assumes on an element $(x, \theta) \in domain\ \mathsf{T} \subset P \times \mathsf{R}^1$ will, consistently, be denoted by $x \mathsf{T} \theta$; a suggested reading is "x transferred by θ". Instead of $(x, \theta) \in domain\ \mathsf{T}$ we will usually, but less correctly, say that $x \mathsf{T} \theta$ is defined.

1.1. Definition. A local dynamical system, or *ld system*, on P is a partial mapping T out of $P \times \mathsf{R}^1$ into P which satisfies the following three conditions:

1. To each $x \in P$ there exist α_x, β_x in $\mathsf{R}^\#$ with $-\infty \leqq \beta_x < 0 < \alpha_x \leqq +\infty$ and such that $x \mathsf{T} \theta$ is defined iff $\beta_x < \theta < \alpha_x$.

2. The initial value property,

$$x \mathsf{T} 0 = x\,, \tag{1}$$

obtains for all $x \in P$.

3. The group property,

$$(x \mathsf{T} \theta_1) \mathsf{T} \theta_2 = x \mathsf{T} (\theta_1 + \theta_2)\,, \tag{2}$$

holds if both $x \mathsf{T} \theta_1$ and also the left or right side of (2) is defined. If $domain\ \mathsf{T} = P \times \mathsf{R}^1$ then T is called a global dynamical system or *gd system.*

For definiteness then a gd system on P is simply a map $\mathsf{T} : P \times \mathsf{R}^1 \to P$ such that (1) and (2) hold for all $x \in P$ and all $\theta_k \in \mathsf{R}^1$. In the formulation of 3, attention is drawn to the precise conditions for the validity of (2); for example, there is a profound difference between ld systems as in 1.1 and systems in which the conditions for (2) are modified to read: (2) holds if all the terms concerned are defined (see III,2.21).

In the general case, a formally weaker variant of the initial value property is sufficient.

1.2. Lemma. Conditions 1–3 are equivalent to 1, 3 and 2′ : T maps onto P.

Proof. Obviously 2 implies 2′. Conversely, assume 1, 2′, 3, and take any $x \in P$; then from 2′, $x = y \mathsf{T} \theta$ for suitable y, θ, and from 3 and 1,

$$x \mathsf{T} 0 = (y \mathsf{T} \theta) \mathsf{T} 0 = y \mathsf{T} (\theta + 0) = y \mathsf{T} \theta = x\,,$$

proving 2.

From 2′ it follows, in particular, that *domain* T is non-void iff P itself is non-void (indeed, the first projection of *domain* T is P). Furthermore, the set P is completely determined by T, $P = image\ \mathsf{T}$; it will be called the *carrier* or *phase space* of T. Each set P is the carrier of at least one gd system, namely the *trivial* system defined by $x \mathsf{T} \theta = x$ for all $x \in P$, $\theta \in \mathsf{R}^1$; it will be shown in 4.5 that, on finite carriers, there is no other ld system except this trivial one. As an example of a non-trivial gd system, take for T the addition operator $+$ on R^1, obtaining the *natural* gd system on R^1. This is an instance of a more general and interesting example which will now be described.

Examples. 1.3.1. Consider an autonomous differential equation

$$\frac{dx}{d\theta} = f(x) \tag{3}$$

on a subset $G \subset \mathsf{R}^n$ $(x \in G,\ \theta \in \mathsf{R}^1,\ f : G \to \mathsf{R}^n$; cf. I,4.3), and assume unicity and local existence of solutions. With (3) one may then associate an ld system T on G defined thus: take any $x \in G$, and let y be the unique generalized solution of (3) which has

$$y(0) = x \tag{4}$$

and maximal open interval $I \subset \mathsf{R}^1$ as domain; then set, for all $\theta \in I$,

$$x \mathsf{T} \theta = y(\theta)\,.$$

Obviously the conditions 1.1.1−2 are satisfied. 3 follows quite easily from autonomness of (3): with the preceding notation, set $I = (\beta, \alpha)$, and according to the definition in I,4.3 one has

$$y(\theta) = y(\theta_1) + \int_{\theta_1}^{\theta} f(y(\lambda))\, d\lambda \quad \text{for} \quad \beta < \theta\,,\ \theta_1 < \alpha\,.$$

Then

$$y(\theta + \theta_1) = y(\theta_1) + \int_{\theta}^{\theta+\theta_1} f(y(\lambda))\, d\lambda = y(\theta_1) + \int_0^{\theta} f(y(\lambda + \theta_1))\, d\lambda$$

for $\beta - \theta_1 < \theta < \alpha - \theta_1$, which may also be interpreted as stating that $y(\theta + \theta_1)$ is another generalized solution of (3); hence

$$(x \mathsf{T} \theta_1) \mathsf{T} \theta_2 = y(\theta_1) \mathsf{T} \theta_2 = y(\theta_2 + \theta_1) = x \mathsf{T} (\theta_2 + \theta_1)$$

as required.

It follows immediately that the ld system thus defined is global iff (3) has extendability of solutions.

1.3.2. As a concrete example, take $G = \mathsf{R}^1$, and for (3) the equation of I,1.11,

$$\frac{dx}{d\theta} = x^2 + 1\,.$$

The solutions were exhibited there; hence one has a closed-form description of the corresponding ld system:

$$x \mathsf{T} \theta = \tan(\theta + \arctan x)$$

with

$$\beta_x = -\tfrac{1}{2}\pi - \arctan x\,, \quad \alpha_x = \tfrac{1}{2}\pi - \arctan x\,.$$

1.3.3. Several examples to be constructed later are connected with the following very special case of 1.3.1: take $G = \mathsf{R}^2$, and for (3) the differential equation (in real coordinates x, y)

$$\frac{\mathrm{d}x}{\mathrm{d}\theta} = 1 \,, \quad \frac{\mathrm{d}y}{\mathrm{d}\theta} = 0 \,.$$

The corresponding ld system is global, and may be described by

$$(x, y) \mathsf{T} \theta = (x + \theta, y)$$

for all $(x, y) \in \mathsf{R}^2$, $\theta \in \mathsf{R}^1$.

1.4. Return to the general case described in 1.1. Analogously to 1.3.1, define *solutions* of an ld system T on P as maps, for each $x \in P$,

$$t_x : (\beta_x, \alpha_x) \to P \,, \quad t_x(\theta) = x \mathsf{T} \theta \,.$$

From 1.1 there then follow the relations

$$t_x(0) = x \,, \quad t_{x\mathsf{T}\theta}(\lambda) = t_x(\lambda + \theta)$$

for all $x \in P$ and suitable θ, $\lambda \in \mathsf{R}^1$. Obviously the set of solutions $\{t_x : x \in P\}$ determines T completely.

Vaguely speaking, solutions are obtained by fixing x in $x \mathsf{T} \theta$. This also suggests fixing θ and allowing x to vary: *translations* of an ld system T on P are defined as partial maps, for each $\theta \in \mathsf{R}^1$,

$$T_\theta : \mathit{domain}\ T_\theta \to P \,, \quad \mathit{domain}\ T_\theta = \{x \in P : x \mathsf{T} \theta \text{ defined}\} \,,$$
$$T_\theta(x) = x \mathsf{T} \theta \,.$$

From 1.1 one then has the relations

$$T_0(x) = x \,, \quad T_{\theta'} T_\theta = T_{\theta' + \theta} \,, \quad T_\theta(x) = t_x(\theta)$$

for all x and suitable θ, $\theta' \in \mathsf{R}^1$. Evidently the system of translations $\{T_\theta \mid \theta \in \mathsf{R}^1\}$ determines T completely.

The α_x, β_x appearing in 1.1.1 will be called *bounds* of T, or *escape times* of x; some of their properties will now be noticed.

1.5. Lemma. If T *is an ld system on* P, *then with the notation of 1.1*

$$\alpha_{x\mathsf{T}\theta} = \alpha_x - \theta \,, \quad \beta_{x\mathsf{T}\theta} = \beta_x - \theta$$

for all $x \in P$ *and* $\beta_x < \theta < \alpha_x$.

Proof. Take any $x \in P$ and then any $\theta \in (\beta_x, \alpha_x)$, so that $x \mathsf{T} \theta$ is defined. According to 1.1.3, $(x \mathsf{T} \theta) \mathsf{T} \lambda$ is defined iff $x \mathsf{T} (\theta + \lambda)$ is defined; using 1.1.1 this implies that, for all $\lambda \in \mathsf{R}^1$, the condition

$$\beta_{x\mathsf{T}\theta} < \lambda < \alpha_{x\mathsf{T}\theta}$$

is equivalent to

$$\beta_x < \theta + \lambda < \alpha_x\,, \quad \text{i.e.} \quad \beta_x - \theta < \lambda < \alpha_x - \theta\,.$$

This proves 1.5.

1.6. Corollary. For an $x \in P$ *let* $\alpha_x < +\infty$ *or* $\beta_x > -\infty$*; then the set*

$$\textit{image}\ t_x = \{x\,\mathsf{T}\,\theta : \beta_x < \theta < \alpha_x\}$$

has continuum-many points.

Proof. If $\alpha_x < +\infty$, say, then all $\alpha_{x\mathsf{T}\theta}$ are distinct for θ varying over (β_x, α_x).

1.7. Corollary. Each ld system on a countable carrier is global.

1.8. Lemma 1.5 may suggest the following procedure for obtaining new ld systems from a given ld system T on P: for each $x \in P$ choose new bounds α'_x, β'_x subject to the conditions

$$\beta_x \leqq \beta'_x < 0 < \alpha'_x \leqq \alpha_x$$

and

$$\alpha'_{x\mathsf{T}\theta} = \alpha'_x - \theta\,, \quad \beta'_{x\mathsf{T}\theta} = \beta'_x - \theta$$

for $\beta'_x < \theta < \alpha'_x$. Then partialize T, defining

$$x\,\mathsf{T}'\,\theta = x\,\mathsf{T}\,\theta \quad \text{for} \quad \beta'_x < \theta < \alpha'_x\,.$$

It is readily verified that the resulting partial map T' is again an ld system on P. A rather surprising result is III,2.2 where it is shown that the only possible choice of new bounds is the trivial one

$$\alpha'_x = \alpha_x\,, \quad \beta'_x = \beta_x \quad \text{for all} \quad x \in P\,.$$

1.9. With a given ld system T on P one may associate another ld system T' on P again by the orientation-changing procedure, setting

$$x\,\mathsf{T}'\,\theta = x\,\mathsf{T}\,(-\theta)$$

for all $x \in P$ and $\beta_x < -\theta < \alpha_x$. Evidently the bounds of T' are

$$\alpha'_x = -\beta_x\,, \quad \beta'_x = -\alpha_x\,.$$

2. SEMI-DYNAMICAL SYSTEMS

Let the operator notation of 1.1 also be employed in describing a concept closely related to ld systems.

2.1. Definition. A local semi-dynamical system (*lsd system*) on P is a partial mapping T out of $P \times \mathsf{R}^+$ into P which satisfies the following conditions.

1. To each $x \in P$ there is an $\alpha_x \in \mathsf{R}^{\#}$ with $0 < \alpha_x \leqq +\infty$ and such that $x \mathsf{T} \theta$ is defined iff $0 \leqq \theta < \alpha_x$;

2 and 3 as in 1.1.

If domain $\mathsf{T} = P \times \mathsf{R}^+$ then T will be termed a global semi-dynamical system (gsd system). The concepts and results of the preceding section (1.9 excepted) have rather obvious counterparts applying to lsd systems. In particular, we may speak of thé carrier of an lsd system, of trivial lsd systems, of natural lsd syxtems on R^1 or R^+ and of +solutions

$$t_x^+ : [0, \alpha_x) \to P\,, \quad t_x^\times(\theta) = x \mathsf{T} \theta\,,$$

of translations $\{T_\theta \mid \theta \in \mathsf{R}^+\}$ and also of bounds α_x. Paralleling 1.5–7:

2.2. Lemma. Let T *be an lsd system on P; then*

$$\alpha_{x+\theta} = \alpha_x - \theta \quad for \quad 0 \leqq \theta < \alpha_x\,. \tag{1}$$

If $\alpha_x < +\infty$ *for some* $x \in P$, *then image* t_x^+ *has continuum-many points. Hence, each lsd system on a countable carrier is global.*

Observe that 2.2 implies that either T is global or $\inf\{\alpha_x : x \in P\} = 0$.

2.3. Assume given an ld system T on P. With T one may then associate, in the obvious manner, an *lsd* system T' on P, obtained by restricting *domain* T to the set

$$(P \times \mathsf{R}^+) \cap domain\, \mathsf{T}\,;$$

this process will be termed semi-dynamical (or lsd) *restriction*. Obviously one may also obtain a second lsd system from T by first changing orientation and then restricting the result.

Now not all lsd systems can be obtained by restricting some ld system (see example 2.8 below). Two necessary conditions for this, -unicity and absence of *s*-points, are exhibited in 2.5; these are interesting on their own and, taken in conjuction, constitute a sufficient condition: 2.9.

2.4. Definition. Let T be an lsd system on *P*. An $x \in P$ is said to be an *s-point* if $x = x' \mathsf{T} \theta$ for no $x' \in P$, $\theta > 0$. The system T is said to possess *-unicity* if

$$x \mathsf{T} \theta = x' \mathsf{T} \theta \quad \text{implies} \quad x = x'\,. \tag{2}$$

As an obvious example, the natural gsd system on R^1 has no *s*-points, that on R^+ has precisely one, the origin; both systems have -unicity. The property of being an *s*-point depends, of course, on the lsd system concerned: thus it would be more proper to say that x is an *s*-point *relative to* T or *of* T or *in* T. However, if there is no doubt which lsd system is meant, the italicized phrase will usually be omitted. It should be emphasized that a similar procedure will be used in analogous situations throughout this book.

2.5. Lemma. If T *is the lsd restriction of an ld system* T' *on the same carrier P, then* T *has -unicity and no s-points.*

Proof. Assume $x_1 \mathsf{T}\, \theta = x_2 \mathsf{T}\, \theta$; then $x_k \mathsf{T}\, \theta = x_k \mathsf{T}'\, \theta$ are defined, and

$$x_1 = (x_1 \mathsf{T}'\, \theta)\, \mathsf{T}'\, (-\theta) = (x_1 \mathsf{T}\, \theta)\, \mathsf{T}'\, (-\theta) = (x_2 \mathsf{T}\, \theta)\, \mathsf{T}'\, (-\theta) = \\ = (x_2 \mathsf{T}'\, \theta)\, \mathsf{T}'\, (-\theta) = x_2 \,.$$

This proves that T has -unicity. Secondly, take any $x \in P$, and then $-\theta \in (\beta'_x, 0)$; hence $\theta > 0$ and $x' = x\, \mathsf{T}'\, (-\theta)$ is defined, whereupon

$$x = x\, \mathsf{T}'\, (-\theta + \theta) = (x\, \mathsf{T}'\, (-\theta))\, \mathsf{T}'\, \theta = x'\, \mathsf{T}\, \theta \,,$$

so that x is not an s-point. This completes the proof.

The following lemma presents two useful reformulations of the -unicity property, formally weaker and stronger respectively.

2.6. Lemma. The -unicity property for an lsd system T *on P is equivalent to either of these conditions:*

1. $x\, \mathsf{T}\, \theta = x'\, \mathsf{T}\, \theta'$ *and* $\theta' \geqq \theta$ *imply* $x = x'\, \mathsf{T}\, (\theta' - \theta)$.

2. *There exists an* $\varepsilon > 0$ *such that, for all* x, x' *in P,*

$$x\, \mathsf{T}\, \theta = x'\, \mathsf{T}\, \theta \quad \textit{and} \quad 0 \leqq \theta \leqq \varepsilon \quad \textit{imply} \quad x = x' \,.$$

Proof. 2.6.1. With $\theta = \theta'$, 1 is (2). Conversely, assume the premises of 1; then

$$x\, \mathsf{T}\, \theta = x'\, \mathsf{T}\, \theta' = x'\, \mathsf{T}\, (\theta' - \theta + \theta) = (x'\, \mathsf{T}\, (\theta' - \theta))\, \mathsf{T}\, \theta \,,$$

so that (2) implies the conclusion of 1. Thus 1 is indeed equivalent with -unicity.

2.6.2. Evidently, -unicity implies 2 with arbitrary ε. Conversely assume 2 and

$$x\, \mathsf{T}\, \theta = x'\, \mathsf{T}\, \theta \,. \tag{3}$$

Then either $\theta \leqq \varepsilon$ and $x = x'$ by assumption; or $\theta > \varepsilon$, and (3) implies

$$(x\, \mathsf{T}\, (\theta - \varepsilon))\, \mathsf{T}\, \varepsilon = (x'\, \mathsf{T}\, (\theta - \varepsilon))\, \mathsf{T}\, \varepsilon \,,$$

whereupon 2 yields

$$x\, \mathsf{T}\, (\theta - \varepsilon) = x'\, \mathsf{T}\, (\theta - \varepsilon) \,.$$

On repeating this process (at most $n + 1$ times, with n the integral part of θ/ε), there results $x = x'$ as asserted. This completes the proof of 2.6.

2.7. Proposition. If T *and* T' *are lsd systems with* $\mathsf{T} \subset \mathsf{T}'$ *(not necessarily on the same carrier) then* T *has -unicity if* T' *does.*

Proof. From $\mathsf{T} \subset \mathsf{T}'$, $x\, \mathsf{T}\, \theta = x'\, \mathsf{T}\, \theta$ implies $x\, \mathsf{T}'\, \theta = x'\, \mathsf{T}'\, \theta$ and hence $x = x'$.

One might express 2.7 as stating that -unicity is hereditary. From 2.5 and 2.7 it follows that an lsd system T has -unicity if there exists an ld system $\mathsf{T}' \supset \mathsf{T}$; the converse assertion will be proved in Chapter V. Since the natural gsd system $+$ on R^+ has an *s*-point, according to 2.5 it is not the semi-dynamical restriction of any ld system; however, there do exist ld systems $\mathsf{T} \supset +$, for example the natural gd system on R'. Proposition 2.7 then implies that, if an lsd system T does not have -unicity, then there cannot exist any ld system $\mathsf{T}' \supset \mathsf{T}$.

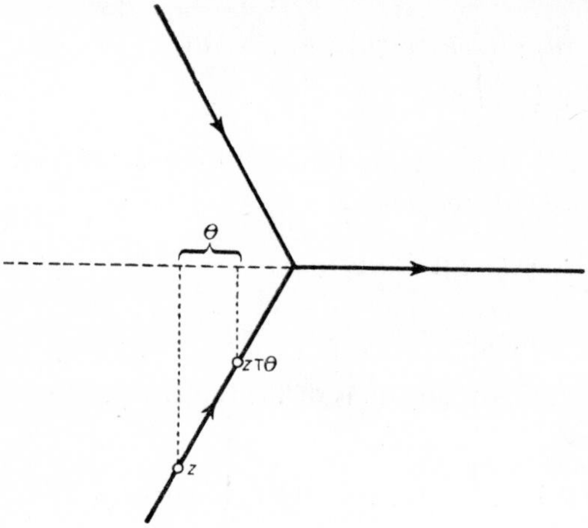

Fig. 1. gsd system without -unicity.

2.8. *Example.* There exist very simple gsd systems without -unicity. To show this let D be the set of all complex $z \in \mathsf{R}^2$ with

$$\operatorname{Arg} z = \tfrac{2}{3}k\pi i \quad (k = 0, \pm 1)$$

or with $z = 0$, i.e. a "triode" (see Fig. 1). For $z \in D$, $\theta \in \mathsf{R}^+$, let $z \,\mathsf{T}\, \theta$ be the point of D obtained by moving z continuously along D and increasing its real part by θ. Obviously this defines a gsd system T on D, and T does not have -unicity:

$$\exp\left(\pm\tfrac{2}{3}\pi i\right) \mathsf{T}\, 1 = 0\,.$$

2.9. *Theorem. An lsd system is the semi-dynamical restriction of some ld system (on the same carrier) iff it has -unicity and no s-points.*

Proof. Necessity of the conditions has been proved in 2.5; assume then that one has an lsd system T on P with -unicity and no *s*-points. Define T' in the possibly obvious manner by setting, for $\theta \in \mathsf{R}^+$,

$$x \,\mathsf{T}'\, \theta = x \,\mathsf{T}\, \theta \quad \text{if} \quad x \,\mathsf{T}\, \theta \text{ is defined} \tag{4}$$

$$x \,\mathsf{T}'\, (-\theta) = y \quad \text{if} \quad x = y \,\mathsf{T}\, \theta\,. \tag{5}$$

From (4) it follows that, if T' is an ld system on P, then T is its lsd restriction. Note that -unicity of T implies that the y in (5) is determined uniquely by x, θ.

Now, if $\theta \leqq 0$ and $x \,\mathsf{T}'\, \theta$ is defined, then also $x \,\mathsf{T}'\, \theta'$ is defined for all θ' with $\theta \leqq \theta' < \alpha_x$, and for all $\theta' < \theta$ sufficiently near to θ; the latter follows directly from the absence of *s*-points. This proves 1.1.1.

Since 1.1.2 coincides with 2.1.2, it only remains to verify the group-property 1.1.3. The proof of this has six cases according to possible combinations of signs of θ_1, θ_2, $\theta_1 + \theta_2$; all of these are elementary exercises in applying (1) and (3). To illustrate this, consider the case $\theta_1 \geqq 0 \geqq \theta_2$, $\theta_1 + \theta_2 \leqq 0$, and assume that $x \mathsf{T}' \theta_1 = x \mathsf{T} \theta_1$ is defined. There are two subcases. If

$$(x \mathsf{T} \theta_1) \mathsf{T}' \theta_2 = y \tag{6}$$

is defined, then from (4) $x \mathsf{T} \theta_1 = y \mathsf{T} (-\theta_2)$; using (3) and $-\theta_2 - \theta_1 \geqq 0$, one concludes that

$$x = y \mathsf{T} (-\theta_2 - \theta_1) ;$$

hence by (4) there is indeed defined

$$x \mathsf{T}' (\theta_2 + \theta_1) = y ,$$

and the required equality follows hence and from (6). The second subcase is that

$$x \mathsf{T}' (\theta_1 + \theta_2) = y \tag{7}$$

is defined; then from (4) $x = y \mathsf{T} (-\theta_1 - \theta_2)$ and, since $x \mathsf{T} \theta_1$ is defined by assumption,

$$x \mathsf{T} \theta_1 = (y \mathsf{T} (-\theta_1 - \theta_2)) \mathsf{T} \theta_1 = y \mathsf{T} (-\theta_2) .$$

Again from (4) there is indeed defined

$$(x \mathsf{T} \theta_1) \mathsf{T}' \theta_2 = y$$

and the required result follows hence and from (7).

The proof of the remaining five cases in 1.1.3 is either similar to the preceding or simpler. This concludes the proof of 2.9.

2.10. The following procedure for obtaining new lsd systems from given ones is suggested by 2.2 and 1.8. Having an lsd system T on P, for each $x \in P$ choose a new escape time α'_x subject to

$$0 < \alpha'_x \leqq \alpha_x ,$$

$$\alpha'_{x \mathsf{T} \theta} = \alpha'_x - \theta \quad \text{for} \quad 0 \leqq \theta \times \alpha'_x ,$$

and then define T' by

$$x \mathsf{T}' \theta = x \mathsf{T} \theta \quad \text{for} \quad 0 \leqq \theta < \alpha'_x . \tag{8}$$

It is easily verified that T' is an lsd system on P, and (8) is then equivalent to $\mathsf{T}' \subset \mathsf{T}$. Thus T' has -unicity if T does (cf. 2.7), and each s-point of T is an s-point of T'. However, new s-points may well appear, so that, in contrast to the situation of ld systems announced in 1.8, the present procedure indeed yields new lsd systems.

2.11. Example. As P, T take R^1 with the natural gsd system $+$, choose new bounds

$$\alpha'_x = +\infty \quad \text{for} \quad x \geqq 0 , \quad \alpha'_x = -x \quad \text{for} \quad x < 0 .$$

On applying 2.10 there results a non-global lsd system T' on R^1 with $\mathsf{T}' \subset +$, with -unicity and with 0 as unique s-point.

2.12. Given an lsd (or ld) system T on P, and also $X \subset P$, $A \subset \mathsf{R}^1$, set

$$X \mathsf{T} A = \{x \mathsf{T} \theta : x \in X, \theta \in A, x \mathsf{T} \theta \text{ defined}\} = image\ \mathsf{T} \mid (X \times A).$$

For $x \in P$ and $\theta \in \mathsf{R}^1$ we then write

$$x \mathsf{T} A = (x) \mathsf{T} A, \quad X \mathsf{T} \theta = X \mathsf{T} (\theta);$$

however, $x \mathsf{T} \theta$ will never be used to replace $(x) \mathsf{T} (\theta)$, since it would then have two meanings. We emphasize that, with this convention, $x \mathsf{T} R^+$ is well defined even if x has finite positive espace time; however $X \mathsf{T} A$ may be void even if both X, A are not.

2.13. Lemma. Let T *be an lsd (or ld) system on* P, $X, X_i \subset P$, $A, A_j \subset \mathsf{R}^1$. *Then the following formulae hold*

$$X \mathsf{T} A = \bigcup_{\theta \in A} (X \mathsf{T} \theta), \quad (\bigcup_i X_i) \mathsf{T} (\bigcup_j A_j) = \bigcup_{i,j} (X_i \mathsf{T} A_j),$$

$$X \mathsf{T} 0 = X, \quad (X \mathsf{T} \mathsf{R}^+) \mathsf{T} \mathsf{R}^+ = X \mathsf{T} \mathsf{R}^+.$$

If T *has -unicity, then for* $\theta \in \mathsf{R}^1$

$$(\bigcap_i X_i) \mathsf{T} \theta = \bigcap_i (X_i \mathsf{T} \theta), \quad (X_1 - X_2) \mathsf{T} \theta = (X_1 \mathsf{T} \theta) - (X_2 \mathsf{T} \theta).$$

As an application of this notation, one has the

2.14. Lemma. If T *is an lsd system on* P, *then the set of s-points is*

$$P - (P \mathsf{T} (0, +\infty));$$

hence P *has no s-points iff* $P \mathsf{T} (0, +\infty) = P$.

Proof. From 2.4 directly, $P \mathsf{T} (0, +\infty)$ is the set of non-s-points.

3. CATEGORIES OF DYNAMICAL SYSTEMS

From now on, the term dynamical system (without further qualification) will occasionally be used as a generic name for ld or lsd systems. Vaguely speaking, a dynamical system defines what may be termed a *dynamical structure* on its carrier set. One may then also speak of dynamical systems (P, T) where P is a set and T a dynamical structure (i.e. an ld or lsd system) on P; however in (P, T) the first term, P, is completely determined by T and hence is superfluous.

Between dynamical systems we may and shall interpret $\mathsf{T} \subset \mathsf{T}'$ as set-inclusion; this relation is then equivalent to T being a partial map of T', and implies $P \subset P'$ for the corresponding carriers (however, $\mathsf{T} \subset \mathsf{T}' \neq \mathsf{T}$ does not exclude $P = P'$, see example 2.11).

3.1. Assume given an indexed system (P_i, T_i) of gd systems for $i \in I$, an index set. Then, on the set-theoretical product $P = \Pi P_i$ as carrier, one may define a gd system T in the natural manner,

$$[x_i] \mathsf{T} \theta = [x_i \mathsf{T}_i \theta] \tag{1}$$

for all $[x_i]_{i \in I} \in P$ and $\theta \in \mathsf{R}^1$. The canonic projection maps p_j (for $j \in I$),

$$p_j : P \to P_j, \quad p_j[x_i] = x_j,$$

then have the property that

$$p_j(x \mathsf{T} \theta) = p_j(x) \mathsf{T}_j \theta \tag{2}$$

for all $x \in P$, $\theta \in \mathsf{R}^1$.

A similar construction may be carried out for gsd systems, by simply replacing R^1 with R^+ throughout. However, in general this is not the case for ld or lsd systems; as will be seen in 3.2, the reason is that, in (1), it is required to have $x_i \mathsf{T}_i \theta$ defined for all $i \in I$.

3.2. Proposition. For $i \in I$ let (P_i, T_i) be lsd systems, $P_i \neq \emptyset$. A necessary and sufficient condition for existence of an lsd system T on $P = \Pi P_i$, such that (2) holds whenever $x \mathsf{T} \theta$ is defined, is that the subset J of those indices $j \in I$ which have T_j non-global be finite. In the positive case T is global iff all T_i are global; T has -unicity if each T_i has -unicity (iff, for finite I); $[x_i] \in P$ is an s-point if some x_i is an s-point for T_i (iff, for finite I). Similar conclusions hold for (P_i, T_i) ld systems.

Proof. 3.2.1. Observe that (2) is equivalent to (1), only requiring that the terms on the left be defined. Hence, if $x = [x_i] \in P$, then

$$\alpha_x \leqq \inf \alpha_{x_i}^{(i)}, \tag{3}$$

where, of course, $\alpha^{(i)}$ denotes the bound in the i-th system T_i.

Now assume there is a countable infinite subset $J \subset I$ such that T_n is non-global for $n \in J$; in P_n there then exists a y_n with finite escape time, and one may define

$$x_n = y_n \mathsf{T}_n \beta_n, \quad \beta_n = \max\left(0, \alpha_{y_n}^{(n)} - \frac{1}{n}\right),$$

so that

$$\alpha_{x_n}^{(n)} \leqq \frac{1}{n}.$$

For $i \notin J$ choose $x_i \in P_i$ arbitrarily, and set $x = [x_i]$. Then from (3)

$$\alpha_x \leqq \alpha_{x_n}^{(n)} \leqq \frac{1}{n} \to 0\,,$$

contradicting 2.1.1.

3.2.2. Secondly, let there exist a finite $J \subset I$ such that, for $i \in I - J$, T_i is global. Define T by (1) whenever $x = [x_i] \in P$ and $0 \leqq \theta < \alpha_x$ where α_x denotes

$$\alpha_x = \inf_I \alpha_{x_i}^{(i)} = \min_J \alpha_{x_j}^{(j)}\,. \tag{4}$$

It is then easily verified that T is an lsd system on P, global iff all T are global. Evidently a similar proof may be carried out for ld systems (P_i, T_i).

3.2.3. There remain the assertions on -unicity and *s*-points. If all T_i have -unicity, then

$$[x_i]\,\mathsf{T}\,\theta = [x_i']\,\mathsf{T}\,\theta$$

in (P, T) implies

$$x_i\,\mathsf{T}_i\,\theta = x_i'\,\mathsf{T}_i\,\theta \quad \text{and} \quad x_i = x_i' \quad \text{for all} \quad i \in I\,;$$

thus T has -unicity.

3.2.4. The converse to 3.2.3. is asserted only for finite index sets I. In this case $\alpha \in \mathsf{R}^{\#}$ defined by

$$\alpha = \inf_{i \in I} \sup_{x \in P_i} \alpha_x^{(i)}$$

is positive, so that one may choose an ε with $0 < \varepsilon < \alpha$. Assuming T has -unicity, we shall verify 2.6.2 for any fixed T_j. Thus, let

$$x_j, x_j' \in P_j\,, \quad 0 \leqq \theta \leqq \varepsilon\,, \quad x_j\,\mathsf{T}_j\,\theta = x_j'\,\mathsf{T}_j\,\theta\,.$$

For each $i \in I$ with $i \neq j$ choose $x_i = x_i'$ such that $x_i\,\mathsf{T}_i\,\theta$ is defined (see definition of α above) and set

$$x = [x_i]\,, \quad x' = [x_i'] \quad \text{in} \quad P\,.$$

Then $x\,\mathsf{T}\,\theta = x'\,\mathsf{T}\,\theta$ in (P, T), yielding $x = x'$ by assumption, and in particular $x_j = x_j'$. Therefore T_j indeed has -unicity.

3.2.5. If $x = [x_i] \in P$ is a non-*s*-point, then, for some $[y_i] \in P$ and $\theta > 0$,

$$x_i = y_i\,\mathsf{T}_i\,\theta \quad \text{for all} \quad i \in I\,,$$

so that all x_i are non-*s*-points.

The converse is asserted only for finite I. Let $x_i \in P_i$ be non-s-points,

$$x_i = y_i \mathsf{T}_i \theta_i, \quad y_i \in P_i, \quad \theta_i > 0;$$

and set

$$\theta = \min \theta_i, \quad z_i = y_i \mathsf{T}_i (\theta_i - \theta).$$

Then $\theta > 0$, $0 \leqq \theta_i - \theta < \theta_i$ (so that z_i is indeed defined), and obviously

$$[z_i] \mathsf{T} \theta = [z_i \mathsf{T}_i \theta] = [y_i \mathsf{T}_i \theta_i] = [x_i],$$

so that $[x_i]$ is a non-s-point. This completes the proof of 3.2.5 and 3.2.

3.3. The dynamical system T thus constructed on $P = \Pi P_i$ will be called the *direct product* of the dynamical systems (P_i, T_i); there are two cases, concerning lsd systems and ld systems, respectively. For definiteness, T exists iff the condition described in 3.2 is satisfied, and it is defined by formula (1) with bounds determined by (4). Concerning unicity of the construction, see III,2.16–17.

Examples. 3.3.1. The gd system on R^2 in example 1.3.3 is the direct product of the natural gd system on R^1 with the trivial gd system on R^1.

3.3.2. There exist gsd systems (P_i, T_i) with -unicity and no s-points such that their direct product does have s-points. To see this, for each positive integer n let T_n be the natural gsd system on $P_n = (0, +\infty)$; then the (P_n, T_n) are as asserted, the direct product exists, but each $[x_n] \in P = \Pi P_n$ with $\inf x_n = 0$ is an s-point. Note that the direct product of the natural ld systems (not lsd) on P_n does not exist, and compare this with 2.9.

3.4. Direct sums of dynamical systems may also be defined. For $i \in I$ let (P_i, T_i) be, for example, ld systems, and for simplicity assume that the system $\{P_i \mid i \in I\}$ is disjoint. On $P = \bigcup P_i$ as carrier, define an ld system T in the obvious manner: for any $x \in P$ there is a unique $i \in I$ with $x \in P_i$, whereupon set

$$x \mathsf{T} \theta = x \mathsf{T}_i \theta$$

iff the latter is defined. The canonic injection maps q_j $(j \in I)$,

$$q_j : P_j \to P, \quad q_j(x) = x,$$

then have the property that

$$q_j(x \mathsf{T}_j \theta) = q_j(x) \mathsf{T} \theta \tag{5}$$

if $x \mathsf{T}_j \theta$ is defined. Evidently, T is global iff all T_i are global.

Similarly, of course, for lsd systems; furthermore, T has -unicity iff all T_i have, and s-points in P_i and P coincide. In either case the dynamical system T will be termed the *direct sum* of the (P_i, T_i). Concerning unicity of the construction see

III,2.18. As an example, take the lsd system T' on R^1 constructed in 2.11; this may be interpreted as the direct sum of an lsd system on $(-\infty, 0)$ with the natural gsd system on R^+.

3.5. Next, consider a dynamical system (P, T), together with an equivalence relation $\sim$ on P which satisfies the *compatibility condition*

$$x_1 \sim x_2 \quad \text{implies} \quad x_1 \,\mathsf{T}\, \theta \sim x_2 \,\mathsf{T}\, \theta$$

whenever both $x_k \,\mathsf{T}\, \theta$ are defined; if furthermore, $x_1 \sim x_2$ implies that $x_1 \,\mathsf{T}\, \theta$ is defined iff $x_2 \,\mathsf{T}\, \theta$ is defined, then $\sim$ will be termed *strongly compatible* with T (an example of this latter situation appears in 4.13; for further applications see IV,3.8).

Evidently, for global dynamical systems compatibility and strong compatibility are equivalent.

To P and $\sim$ there corresponds the set $P' = P/\sim$ of equivalence classes modulo $\sim$; with this set P' and the given dynamical system T on P there will be associated a partial map out of $P' \times \mathsf{R}^1$ into P', to be denoted by T' and defined in the following manner.

3.5.1. Take any $x \in P'$ and $\theta \in \mathsf{R}^1$; then choose, if possible, $x_k \in P$ and $\theta_k \in \mathsf{R}^1$ for $1 \leqq k \leqq n \in \mathsf{C}^+$ such that

$$\sum_1^n \theta_k = \theta\,, \quad x_1 = x\,, \quad x_{k+1} \sim x_k \,\mathsf{T}\, \theta_k \quad \text{for} \quad 1 \leqq k < n\,;$$

finally, let y' be the equivalence class modulo $\sim$ containing the element $x_n \,\mathsf{T}\, \theta_n$ (this implies the requirement that all $x_k \,\mathsf{T}\, \theta_k$ be defined). The value $x' \,\mathsf{T}'\, \theta$ of T' at (x', θ) is then defined as the element $y' \in P'$ just obtained; if no such x_k, θ_k exist then $x' \,\mathsf{T}'\, \theta$ will be left undefined. To verify that T' is indeed a partial map it is necessary to show that y' is independent of the particular choice of the sequences $\{x_k\}$, $\{\theta_k\}$ employed. This will be performed in the following subitems.

3.5.2. First show that for fixed $\{\theta_k\}_1^n$, the y' is independent of the $\{x_k\}$. Thus, let also $u_1 = x$, $u_{k+1} \sim u_k \,\mathsf{T}\, \theta_k$; then by induction, from compatibility and $u_1 = x_1$, there is $u_k \sim x_k$, so that $u_n \,\mathsf{T}\, \theta_n \sim x_n \,\mathsf{T}\, \theta_n$ are indeed contained in the same equivalence class y'.

3.5.3. To prove independence on the $\{\theta_k\}$, it is useful to reformulate the construction slightly. Assume that T is an lsd system (and hence $\theta, \theta_k \in \mathsf{R}^+$). Set $\xi_0 = 0$, $\xi_k = \sum_1^k \theta_j$. Then

$$0 = \xi_0 \leqq \xi_1 \leqq \ldots \leqq \xi_n = \theta \tag{6}$$

is a decomposition of the interval $[0, \theta]$, and the construction may be formulated in terms of these ξ_k: $\theta_k = \xi_k - \xi_{k-1}$,

$$x_{k+1} \sim x_k \mathsf{T} (\xi_k - \xi_{k-1}), \quad x_n \mathsf{T} (\xi_n - \xi_{n-1}) \in y' .$$

3.5.4. Now introduce one further dividing point ξ in the decomposition (6), say within $[\xi_{k-1}, \xi_k]$. Then, obviously, the element $u = x_k \mathsf{T} (\xi - \xi_{k-1})$ is defined and has

$$u \mathsf{T} (\xi_k - \xi) = x_k \mathsf{T} (\xi_k - \xi_{k-1}) \sim x_{k+1} .$$

Hence the sequence $x_1, \ldots, x_k, u, x_{k+1}, \ldots, x_n$ and decomposition $\xi_0, \xi_1, \ldots, \xi_{k-1}, \xi, \xi_k, \ldots, \xi_n$ may also be used to define $x' \mathsf{T}' \theta$ and of course yield the same $y' \ni \ni x_n \mathsf{T} (\xi_n - \xi_{n-1})$. Necessarily, a like result obtains for several dividing points, for example introduced successively. Thus the construction of $x' \mathsf{T}' \theta$ is invariant under subdivisious of the decomposition (6).

3.5.5. The proof of independence (for lsd systems) is now almost complete: it suffices to take a common subdivision of two decompositions of $[0, \theta]$, and then apply first 3.5.4 and then 3.5.2. As concerns ld systems, it is easily checked that the same proof may be carried out, replacing R^+ by R^1 throughout (even though one no longer has $\xi_{k-1} \leqq \xi_k$ in (6)).

3.5.6. It should be emphasized that if $\sim$ is strongly compatible with T, then the construction is considerably simpler, since it terminates at the first step. Indeed

$$x_2 \sim x_1 \mathsf{T} \theta_1 = x \mathsf{T} \theta_1 , \quad x_2 \mathsf{T} \theta_2 \text{ defined}$$

now imply that $x \mathsf{T} (\theta_1 + \theta_2)$ is also defined, and by induction, that $x \mathsf{T} \theta = x \mathsf{T} \sum \theta_k$ is defined. In this case then, y' is directly the equivalence class containing $x \mathsf{T} \theta$.

Concerning this construction one has the following result.

3.6. Proposition. Let T be an lsd system on P, and $\sim$ an equivalence relation on P compatible with T. Then the partial map T' constructed in 3,5 is an lsd system on $P' = P/\sim$, and if $e : P \to P'$ is the canonic quotient mapping (i.e. $x \in e(x) \in P'$), then

$$e(x \mathsf{T} \theta) = e(x) \mathsf{T}' \theta \tag{7}$$

if $x \mathsf{T} \theta$ is defined. If T is global then so is T'; if $e(x)$ is an s-point then so is x. Similar results hold for ld systems.

Proof. 3.6.1. First consider lsd systems. Obviously $x' \mathsf{T}' 0 = x'$ for all $x' \in P'$, the initial value property. Now take any $x' \in P'$, $\theta \in \mathsf{R}^+$ with $x' \mathsf{T}' \theta$ defined, and then x_k, θ_k as in 3.5.1; thus $x' \mathsf{T}' \theta = e(x_n \mathsf{T} \theta_n)$. For any $\lambda \in [0, \theta]$ there is $\theta_k \leqq \lambda \leqq \theta_{k+1}$ for some k, and hence $x' \mathsf{T}' \lambda = e(x_k \mathsf{T} \lambda)$ is defined. Also, since $x_n \mathsf{T} \theta_n$ is defined, so is $x_n \mathsf{T} (\theta_n + \lambda)$ for small $\lambda > 0$, and therefore $x' \mathsf{T}' (\theta + \lambda) = e(x_n \mathsf{T} (\theta_n + \lambda))$

is defined for small $\lambda > 0$. This shows that $x' \mathsf{T}' \theta$ is defined for each $x' \in P'$ and for θ in some right-open neighbourhood of 0 in R^+ as required in 2.1.1. For ld systems the proof is similar.

3.6.2. To show that T' is a dynamical system, it remains to verify the group property; again consider lsd systems first. Take any $x' \in P'$ and $\theta, \theta' \in \mathsf{R}^+$. The two required assertions are established in the following two subitems.

3.6.3. If $x' \mathsf{T}' (\theta + \theta')$ is defined, there exist x_k, θ_k with

$$\sum \theta_k = \theta + \theta' , \quad x_1 = x , \quad x_{k+1} \sim x_k \mathsf{T} \theta_k .$$

Since $0 \leqq \theta \leqq \theta + \theta'$, there is $\theta_k \leqq \theta \leqq \theta_{k+1}$ for some k. Then $x' \mathsf{T}' \theta = e(x_k \mathsf{T} \theta)$ is defined; moreover, $x_k \mathsf{T} \theta, x_{k+1}, \ldots, x_n$ and $\theta_k - \theta, \theta_{k+1}, \ldots, \theta_n$ serve to define $(x' \mathsf{T}' \theta) \mathsf{T} \theta'$ and prove that

$$x' \mathsf{T}' (\theta + \theta') = (x' \mathsf{T}' \theta) \mathsf{T}' \theta' . \tag{8}$$

3.6.4. Conversely, assume that both $x' \mathsf{T}' \theta$ and $(x' \mathsf{T}' \theta) \mathsf{T}' \theta'$ are defined, say via x_k, θ_k and u_k, λ_k respectively. Thus

$$\sum_1^n \theta_k = \theta , \quad x_1 = x , \quad x_{k+1} \sim x_k \mathsf{T} \theta_k ,$$

$$\sum_1^n \lambda_k = \theta' , \quad u_1 \sim x_n \mathsf{T} \theta_n , \quad u_{k+1} \sim u_k \mathsf{T} \lambda_k .$$

Then obviously $x_1, \ldots, x_n, u_1, \ldots, u_m$ and $\theta_1, \ldots, \theta_n, \lambda_1, \ldots, \lambda_m$ serve to define $x' \mathsf{T}$ $\mathsf{T}' (\theta + \theta')$ and verify (8). Again, for ld systems the working of the proof is similar.

3.6.5. Relation (6) follows directly from the construction 3.5.1, and the assertions on globality and *s*-points are trivial consequences. This completes the proof of 3.6.

The dynamical system (P', T') constructed in 3.5 will be called the *factor dynamical system* of (P, T) modulo $\sim$. From (6) it follows that

$$\alpha_x \leqq \alpha'_{e(x)}$$

for all $x \in P$; and from 3.5 one has that an equivalence relation $\sim$ compatible with an lsd system T is strongly compatible with T iff

$$\alpha_x = \alpha'_{e(x)} \quad \text{for} \quad x \in P ,$$

(and similarly for ld systems).

Examples. 3.7.1. The factor system of a non-global dynamical system may well be global. Thus, let

$$P = \{(x, y) \in \mathsf{R}^2 : x < 0 \text{ or } y \geqq 0\} ;$$

for $(x, y) \in P$, $\theta \in \mathsf{R}^1$ let

$$(x, y) \mathsf{T} \theta = (x + \theta, y) \tag{9}$$

if the latter point is in P (cf. 1.3.3); between points of P let $(x, y) \sim (x', y')$ iff

$$x = x', \quad |y| = |y'| .$$

Then T is a non-global ld system on P,

$$\alpha_{(x,y)} = +\infty \quad \text{for} \quad y \geqq 0, \quad \alpha_{(x,y)} = -x \quad \text{for} \quad y < 0 ;$$

also, $\sim$ is an equivalence relation on P compatible with T; it is readily verified that the corresponding factor ld system is global (indeed, it is elegantly interpreted in the closed upper half-plane).

3.7.2. In the process of factorization the -unicity property may be lost. Thus, let

$$P = \{(x, y) \in \mathsf{R}^2 : x \geqq 0 \text{ or } y \neq 0\} ,$$

and for $(x, y) \in P$, $\theta \in \mathsf{R}^+$ define T by (9); between elements of P let $(x, y) \sim (x', y')$ iff

$$x = x', \quad (x, \lambda y + (1 - \lambda) y') \in P \quad \text{for all} \quad \lambda \in [0, 1] .$$

Then T is a gsd system on P with -unicity, and $\sim$ is an equivalence relation compatible with T. However, the factor system of (P, T) modulo $\sim$ does not have -unicity; indeed, it is isomorphic (in the obvious sense, see also 3.9) to the dynamical system of 2.8.

3.7.3. It is easily shown that, in the situation of 3.6, T' has -unicity iff

$$x \mathsf{T} \theta \sim y \mathsf{T} \theta \quad \text{implies} \quad x \sim y .$$

In particular, the factor system may have -unicity even if the original system does not; as an example, in 2.8 identify points with equal real parts, obtaining R^1 with the natural gsd system.

3.8. If in 3.5 T is a gd system, then

$$x' \mathsf{T}' \theta = x' \mathsf{T} \theta ,$$

with the right side interpreted in the sense of 2.12 (observe that both sides denote subsets of P). To see this, note that

$$x' \mathsf{T}' \theta \subset x' \mathsf{T} \theta$$

by definition; conversely, $y \sim x \mathsf{T} \theta$ implies $y \mathsf{T} (-\theta) \sim x$, so that $y \mathsf{T} (-\theta) \in x'$,

$$y = (y \mathsf{T} (-\theta)) \mathsf{T} \theta \in x' \mathsf{T} \theta .$$

3.9. We have already used the phrase dynamical structure on a set; this recalls the situation of familiar concrete categories whose objects are sets endowed with structures of special types. In this connection, the following definition is quite natural.

The *category* $\mathscr{D}_l$ *of ld systems* has as objects couples (P, T) with P a set and T an ld system on P; and its morphisms

$$f : (P, \mathsf{T}) \to (P', \mathsf{T}')$$

correspond to those set-theoretical maps $g : P \to P'$ which commute with the given dynamical operators in the sense that

$$g(x \mathsf{T} \theta) = g(x) \mathsf{T}' \theta \tag{10}$$

whenever $x \mathsf{T} \theta$ is defined. This implies, of course, that

$$\beta'_{g(x)} \leqq \beta_x < 0 < \alpha_x \leqq \alpha'_{g(x)} \quad \text{for} \quad x \in P .$$

As in most concrete categories, several morphisms f may correspond to the same map g; nevertheless, adopting the usual convention, the same letter will be used to denote both f and g. For example, let $+$ be the natural gd system in R^1, T_2 the gd system in the complex plane R^2 defined by

$$z \mathsf{T}_2 \theta = ze^{i\theta} \quad \text{for} \quad z \in \mathsf{R}^2 , \quad \theta \in \mathsf{R}^1 ,$$

and T_1 the gd system defined by the same formula on the unit circle S^1 of R^2. Then the map $g : \mathsf{R}^1 \to \mathsf{S}^1$, $g(x) = e^{ix}$, induces distinct morphisms

$$f_1 : (\mathsf{R}^1, +) \to (\mathsf{S}^1, \mathsf{T}_1) , \quad f_2 : (\mathsf{R}^1, +) \to (\mathsf{R}^2, \mathsf{T}^2) ;$$

with our convention both f_1 and f_2 may be denoted by g.

There is an obvious subcategory $\mathscr{D}_g$ of gd systems; analogous definitions introduce the following categories

$\mathscr{D}_{sl}$ ... lsd systems,	$\mathscr{D}_{sg}$ ... gsd systems,
$\mathscr{D}_{slu}$ ... lsd systems with -unicity,	$\mathscr{D}_{sgu}$ ... gsd systems with -unicity,
$\mathscr{D}_l(P)$ ... ld systems on P,	$\mathscr{D}_g(P)$ ... gd systems on P,

etc., and there are the obvious inclusions between these categories.

3.10. The orientation changing procedure of 1.9 is better described as the action of a covariant functor $\mathscr{D}_l \to \mathscr{D}_l$. Probably this is quite obvious; however, the explicit description below may serve as a model for analogous constructions later, which will be merely suggested and not performed in detail.

The orientation changing functor C assigns, to each object (P, T) in $\mathscr{D}_l$ the object $(P^C, \mathsf{T}^C) \in \mathscr{D}_l$ defined as follows:

$$P^C = P\,, \quad x\,\mathsf{T}^C\theta = x\,\mathsf{T}\,(-\theta)$$

iff $x\,\mathsf{T}\,(-\theta)$ is defined; and to each morphism $f : (P, \mathsf{T}) \to (P', \mathsf{T}')$ in $\mathscr{D}_l$, C assigns the morphism

$$f^C : (P^C, \mathsf{T}^C) \to (P'^C, \mathsf{T}'^C)$$

in $\mathscr{D}_l$ again, defined by the same set-theoretical map $f : P \to P'$:

$$f(x\,\mathsf{T}\,\theta) = f(x)\,\mathsf{T}'\,\theta\,,$$
$$f(x\,\mathsf{T}^C\,\theta) = f(x\,\mathsf{T}\,(-\theta)) = f(x)\,\mathsf{T}'\,(-\theta) = f(x)\,\mathsf{T}'^C\,\theta\,.$$

Obviously $C^2 = CC$ is the identity functor of $\mathscr{D}_l$; and C induces functors $C_g : \mathscr{D}_g \to \to \mathscr{D}_g$ and $\mathscr{D}_l(P) \to \mathscr{D}_l(P)$, etc., which commute with the inclusion functors of these categories; for example

$$\begin{array}{ccc} \mathscr{D}_g & \subset & \mathscr{D}_l \\ {\scriptstyle C_g}\downarrow & & \downarrow{\scriptstyle C} \\ \mathscr{D}_g & \subset & \mathscr{D}_l \end{array}$$

3.11. Remark. Any endomorphism h of the additive group of R^1 induces a covariant functor $\mathscr{D}_l \to \mathscr{D}_l$ by assigning to any ld system T on P the ld system T^h on P again, defined by

$$x\,\mathsf{T}^h\,\theta = x\,\mathsf{T}\,h(\theta)\,.$$

The choice $h(\theta) = -\theta$ for all $\theta \in \mathsf{R}^1$ yields the orientation changing functor; the choice $h(\theta) \equiv 0$ yields the trivializing functor.

3.12. The procedure of taking the semi-dynamical restriction of an ld system (cf. 2.3) is better described as the action of an obvious covariant functor $G : \mathscr{D}_l \to \mathscr{D}_{sl}$. One of the assertions in 2.5 may then be interpreted as a factorization of G, with commutativity as shown in the diagram

$$\begin{array}{ccc} & \mathscr{D}_l & \\ {\scriptstyle R}\swarrow & & \searrow{\scriptstyle G} \\ \mathscr{D}_{slu} & \subset & \mathscr{D}_{sl} \end{array}$$

Again, G induces functors on $\mathscr{D}_g$, $\mathscr{D}_l(P)$, etc.

3.13. Equations (2), (5), (6) are typical instances of the morphism condition (10). Furthermore, 3.2 may be interpreted, in category-theoretical terms (Kurosh, 1960) as the construction of direct joins in $\mathscr{D}_g$, $\mathscr{D}_{sg}$, $\mathscr{D}_l$, $\mathscr{D}_{sl}$; 3.4 as the construction of free sums in $\mathscr{D}_g$, etc.; 3.6 as the description of factor objects in $\mathscr{D}_g$, etc. There arises naturally the question as to subobjects of a given object in these categories; a characterization of subobjects in $\mathscr{D}_l$ is given in III,1.13.

3.14. Remark. Many properties of ld systems need be proved or defined only for lsd systems, and then transferred back via the restriction functor (and possibly the orientation changing functor). Thus 1.5 follows from 2.2 in this manner; or in Section 5, +invariance is defined for lsd systems, and then used within ld systems. This procedure will often be used without express reference; thus we might formulate 2.5 by saying that each ld system has -unicity and no s-points.

The following lemma is immediate.

3.15. Lemma. Let $f:(P,\mathsf{T}) \to (P',\mathsf{T}')$ be a morphism in $\mathscr{D}_{sl}$; if x is a non-s-point in (P,T), then $f(x)$ is a non-s-point in (P',T').

Proof. Apply 2.14 and

$$f(P\,\mathsf{T}\,(0,+\infty)) = f(P)\,\mathsf{T}'\,(0,+\infty) \subset P'\,\mathsf{T}'\,(0,+\infty)\,.$$

3.16. Assume given an lsd system T on P, a set P', and a $1-1$ mapping onto $f:P\to P'$. Then T and f induce an lsd system T' on P', defined in the obvious manner; for $x'\in P'$, $\theta\in\mathsf{R}^+$ set

$$x'\,\mathsf{T}'\,\theta = f(f^{-1}(x')\,\mathsf{T}\,\theta)\,, \tag{11}$$

iff the latter term is defined. It follows that f induces an isomorphism in $\mathscr{D}_{sl}$,

$$f:(P,\mathsf{T})\to(P',\mathsf{T}')\,,$$

so that in (P',T') there are reproduced all the dynamical properties of (P,T). Of course, similar remarks apply to ld systems.

This situation is much more general than may appear at first glance. Thus, consider the natural ld system $+$ on the open-segment $(-\frac{1}{2}\pi,\frac{1}{2}\pi)\subset\mathsf{R}^1$, i.e. the system defined by

$$x+\theta \quad \text{if} \quad |x|<\tfrac{1}{2}\pi \quad \text{and} \quad |x+\theta|<\tfrac{1}{2}\pi\,;$$

and also the map tan: $(-\frac{1}{2}\pi,\frac{1}{2}\pi)\to\mathsf{R}^1$. Then the induced ld system on R^1 is that described in 1.3.2, and (11) corresponds to the identity exhibited there.

3.17. Assume given an ld system T on P and a mapping $f:P\to P$ such that $f:(P,\mathsf{T})\to(P,\mathsf{T})$ in $\mathscr{D}_{sl}$, i.e. such that

$$f(x\,\mathsf{T}\,\theta) = f(x)\,\mathsf{T}\,\theta$$

whenever $x\,\mathsf{T}\,\theta$ is defined; then T will be termed *invariant* relative to f. In the special case that f is also a symmetry of P in the sense that $f^2 = ff$ is the identity map of P, then T will be termed *symmetric* relative to f.

If T is an ld system on P, T' is obtained by orientation change, and if $f : (P, \mathsf{T}) \to$ $\to (P, \mathsf{T}')$ in $\mathscr{D}_l$, i.e. if

$$f(x \mathsf{T} \theta) = f(x) \mathsf{T} (-\theta)$$

whenever $x \mathsf{T} \theta$ is defined, then T will be termed *anti-invariant* relative to f, and *anti-symmetric* relative to f if f is a symmetry of P.

3.18. Example. It may be remarked that the situation just described is of not infrequent occurence (for examples see I,5); as for anti-symmetry, consider the differential equation of a conservative system with n degrees of freedom,

$$\frac{d^2x}{d\theta^2} = g(x),$$

($x \in \mathsf{R}^n$, $\theta \in \mathsf{R}^1$, $g : \mathsf{R}^n \to \mathsf{R}^n$, compare with I,2.2). The reduction to a first-order equation is

$$\frac{dx}{d\theta} = y, \quad \frac{dy}{d\theta} = g(x), \tag{12}$$

an autonomous system in R^{2n}; and the corresponding ld system defined (as in 1.3.1) is anti-symmetric relative to the map $f : \mathsf{R}^{2n} \to \mathsf{R}^{2n}$ defined by

$$f(x, y) = (x, -y).$$

To see this it is only needed to verify that, if $(x(\theta), y(\theta))$ is a solution of (12), then so is

$$(x(-\theta), -y(-\theta)).$$

3.19. Lemma. Let T *be an lsd system on* P, F *a set of* $1-1$ *maps onto* $P \to P$, *and let* T *be invariant relative to all* $f \in F$. *Denote by* F_1 *the least group generated by maps of* F *with map composition as the group operation, and by* $\sim$ *the relation defined on* P *by*

$$x \sim y \quad \textit{iff} \quad f(x) = y \quad \textit{for some} \quad f \in F_1.$$

Then $\sim$ *is an equivalence relation on* P *compatible with* T.

Proof. Since F_1 is a group, $\sim$ is indeed an equivalence relation on P. If $f(x) = y$ for some $x, y \in P$, $f \in F$, then from the assumption that T is invariant relative to f,

$$f(x \mathsf{T} \theta) = f(x) \mathsf{T} \theta = y \mathsf{T} \theta$$

if $x \mathsf{T} \theta$ is defined; in particular, $y \mathsf{T} \theta$ is defined and $x \mathsf{T} \theta \sim y \mathsf{T} \theta$. A similar result holds for f^{-1}, and for compositions of finite systems of such maps, i.e. for all $f \in F_1$. Hence $\sim$ is compatible with T as asserted.

Observe that if in 3.19 F has the property that $f^{-1} \in F$ whenever $f \in F$, then $\sim$ is strongly compatible with T.

3.20. Example. In example 1.3.1, take $G = \mathsf{R}^n$ and assume that $f : \mathsf{R}^n \to \mathsf{R}^n$ is periodic,

$$f(x + c) = f(x) \quad \text{for all} \quad x \in \mathsf{R}^n$$

and some $c \neq 0$ in R^n. Define $g : \mathsf{R}^n \to \mathsf{R}^n$ by

$$g(x) = x + c\,;$$

then g is $1-1$ onto, and the ld system corresponding to the differential equation

$$\frac{dx}{d\theta} = f(x)$$

is invariant relative to g (verify that $x(\theta) + c$ is a solution whenever $x(\theta)$ is). In applying 3.19, let F consist of g alone, so that $F_1 = \{g^m : m \text{ integer}\}$; hence $\sim$ may be described directly,

$$x \sim y \quad \text{iff} \quad x = y + mc$$

for some integer m. One may then apply 3.6 and obtain the factor ld system on $\mathsf{R}^n/\!\sim$; in the special case that $n = 2$ and $c = (0, 2\pi)$ there results the situation of I,3.2, and $\mathsf{R}^2/\!\sim$ is the 2-cylinder $\mathsf{R}^1 \times \mathsf{S}^1$.

4. TRAJECTORIES, CYCLES, CRITICAL POINTS

Recall the notation of 2.12, and assume given an ld system T on P and a point $x \in P$. The sets

$$x\,\mathsf{T}\,\mathsf{R}^1, \quad x\,\mathsf{T}\,\mathsf{R}^+, \quad x\,\mathsf{T}\,\mathsf{R}^- \tag{1}$$

are called, respectively, a *trajectory*, a *+trajectory* and a *−trajectory* (semi-trajectory is a generic name for the latter two). The following assertion is then proved easily.

4.1. Lemma. Let T *be an ld system on* P*; then*

$$x\,\mathsf{T}\,\mathsf{R}^1 = (x\,\mathsf{T}\,\mathsf{R}^+) \cup (x\,\mathsf{T}\,\mathsf{R}^-)$$

for $x \in P$*; also,* $y \in x\,\mathsf{T}\,\mathsf{R}^+$ *implies*

$$x\,\mathsf{T}\,\mathsf{R}^+ \supset y\,\mathsf{T}\,\mathsf{R}^+, \quad x\,\mathsf{T}\,\mathsf{R}^- \subset y\,\mathsf{T}\,\mathsf{R}^-, \quad x\,\mathsf{T}\,\mathsf{R}^1 = y\,\mathsf{T}\,\mathsf{R}^1\,;$$

in particular, distinct trajectories are disjoint.

Now attempt to obtain a similar result for lsd systems T; the sets (1) are well defined, but for example $x\,\mathsf{T}\,\mathsf{R}^- = x$. In some situations, the role of $x\,\mathsf{T}\,\mathsf{R}^-$ may be taken over by the set $\{u : x \in u\,\mathsf{T}\,\mathsf{R}^+\}$ (also see Section 5):

4.2. Lemma. Let T *be an lsd system on* P, $x \in P$, $y \in x\,\mathsf{T}\,\mathsf{R}^+$; *then*

$$x\,\mathsf{T}\,\mathsf{R}^+ \supset y\,\mathsf{T}\,\mathsf{R}^+\,, \quad \{u : x \in u\,\mathsf{T}\,\mathsf{R}^+\} \subset \{u : y \in u\,\mathsf{T}\,\mathsf{R}^+\}\,.$$

If T *is the semi-dynamical restriction of an ld system* T' *on* P, *then*

$$x\,\mathsf{T}\,\mathsf{R}^+ = x\,\mathsf{T}'\,\mathsf{R}^+\,, \quad \{u : x \in u\,\mathsf{T}\,\mathsf{R}^+\} = x\,\mathsf{T}'\,\mathsf{R}^-\,.$$

A basic result of this section is the following

4.3. Proposition. Let T *be an lsd system on* P. *If some* $x \in P$ *has the property that* $x\,\mathsf{T}\,\theta = x\,\mathsf{T}\,\theta'$ *for some* $\theta < \theta'$ *in* R^+, *then* $\alpha_x = +\infty$ *and, with* $\tau = \theta' - \theta$,

$$x\,\mathsf{T}\,\lambda = x\,\mathsf{T}\,(\lambda + n\tau) \tag{2}$$

for all $\lambda \geqq \theta$ *and all integers* $n \geqq 0$. *If* T *has -unicity then* $\lambda \geqq \theta$ *may be replaced by* $\lambda \in \mathsf{R}^+$. *If* T *is an ld system then also* $\beta_x = -\infty$ *and* (2) *holds for all* $\lambda \in \mathsf{R}^1$ *and all integers* n.

Proof. From 2.2,

$$\alpha_x - \theta = \alpha_{x\,\mathsf{T}\,\theta} = \alpha_{x\,\mathsf{T}\,\theta'} = \alpha_x - \theta'\,,$$

so that $\theta \neq \theta'$ implies $\alpha_x = +\infty$. For $\lambda \geqq \theta$,

$$x\,\mathsf{T}\,\lambda = (x\,\mathsf{T}\,\theta)\,\mathsf{T}\,(\lambda - \theta) = (x\,\mathsf{T}\,\theta')\,\mathsf{T}\,(\lambda - \theta) = x\,\mathsf{T}\,(\lambda + \tau)\,;$$

also $\lambda + \tau \geqq \lambda \geqq \theta$, so that

$$x\,\mathsf{T}\,\lambda = x\,\mathsf{T}\,(\lambda + \tau) = x\,\mathsf{T}\,(\lambda + \tau + \tau) = x\,\mathsf{T}\,(\lambda + 2\tau)\,,$$

and (2) by induction on n.

If T has -unicity then from (2) with for example $\lambda = \theta$,

$$x\,\mathsf{T}\,\theta = x\,\mathsf{T}\,(\theta + n\tau) = (x\,\mathsf{T}\,n\tau)\,\mathsf{T}\,\theta\,,$$

there follows $x = x\,\mathsf{T}\,n\tau$, and hence (2) for all $\lambda \geqq 0$ and integers $n \geqq 0$. If T is an ld system then from $x = x\,\mathsf{T}\,\tau$ there follows

$$x\,\mathsf{T}\,(-\tau) = x\,,$$

and all the conclusions follow by change of orientation.

4.4. Corollary. Let T *be an lsd system with -unicity on* P, $x \in P$. *The set of solutions* $\theta \in \mathsf{R}^+$ *of the equation*

$$x = x\,\mathsf{T}\,\theta \tag{3}$$

is the trace on R^+ *of a subgroup* G *of the additive group* R^1; *and then, for each* $y \in x\,\mathsf{T}\,\mathsf{R}^+$, *the set of solutions* $\theta \in \mathsf{R}^+$ *of* $y = x\,\mathsf{T}\,\theta$ *is the trace on* R^+ *of a coset of* G.

(Hint: 2.6.1.) For purposes of reference, the subgroup G appearing in 4.4 will be termed the *null-group* of x; obviously

$$G = \{\pm\theta : x = x \mathsf{T} \theta\} .$$

Now, considerable information is available concerning subgroups G of R^1; for example G is either trivial or cyclic or dense in R^1, and this forms the basis of the classification of 4.6. Furthermore, if $G \neq \mathsf{R}^1$ then R^1/G is infinite (cf. Kurosh, 1953, § 23), so that

4.5. Corollary. On a finite carrier, every lsd system with -unicity is trivial.

4.6. Definition. Let T be an lsd system on P and $x \in P$. Then there obtains precisely one of the following cases.

1. Equation (3) holds for all $\theta \in \mathsf{R}^+$; $x = x \mathsf{T} \mathsf{R}^+$ is then called a *critical point.*

2. x is not critical but (3) holds for arbitrarily small $\theta > 0$; x will then be termed a *feebly critical point.*

3. There exists a least positive solution $\lambda \in \mathsf{R}^+$ of (3); $x \mathsf{T} \mathsf{R}^+$ is then called a *cycle* with (primitive) *period* λ.

4. There is no positive solution of (3); $x \mathsf{T} \mathsf{R}^+$ is then called a *non-cyclic +trajectory.*

The concepts of trajectories, cycles and critical points (in another terminology – trajectories, periodic trajectories and singular points) is possibly familiar from differential equation theory (for example, Coddington and Levinson, 1955); that of feebly critical points definitely is not. In the following chapter, some effort will be devoted to showing that, under very reasonable conditions on P and T, there are no feebly critical points in P. However, an example exhibiting feebly critical points would be in place here.

4.7. Example. Take R^1 with its natural gd system, and any non-trivial dense subgroup G of the additive group of R^1; for example, the set of all rational multiplies of a fixed irrational in R^1. Let $\sim$ be the equivalence relation used to define the quotient group R^1/G, i.e.

$$x \sim y \quad \text{iff} \quad x - y \in G ;$$

obviously $\sim$ is compatible with the gd system $+$ on R^1. The construction 3.5 then yields a gd system T on R^1/G with all points feebly critical. Indeed, each $x' \in \mathsf{R}^1/G$ is a coset, $x' = x + G$ for some $x \in \mathsf{R}^1$, and $x' \mathsf{T} \theta$ is then defined as the coset $x + \theta + {} + G$ (cf. 3.8); then $x' = x' \mathsf{T} \theta$ iff $\theta \in G$, since this is equivalent to

$$x + \theta + G = x + G .$$

4.8. Properly speaking, definition 4.6 is intended for use on lsd systems having -unicity. For general lsd systems, the case included under 4.6.4 falls naturally into the following; either there is a $\lambda > 0$ such that $x \mathsf{T} \lambda$ is critical or feebly critical or on a cycle, or $x \mathsf{T} \theta = x \mathsf{T} \theta'$ for no $\theta \neq \theta'$ in R^+. However, our main interest is in lsd systems with -unicity, so that this suggestion will not be followed through there.

4.9. *Proposition. Let* T *be an lsd system with -unicity on* P, $x \in P$; *set*

$$T = \{u : x \in u \mathsf{T} \mathsf{R}^+\} \cup (x \mathsf{T} \mathsf{R}^+),$$

and consider the classification of 4.6.

In case 1: $T = (x)$.

In case 2: $T = x \mathsf{T} \mathsf{R}^+ = x \mathsf{T} [0, \varepsilon]$ *for any* $\varepsilon > 0$, *and* T *consists of feebly critical points; for each* $y \in T$, *the set of solutions* $\theta \in \mathsf{R}^+$ *of* $y = x \mathsf{T} \theta$ *is dense in* R^+, *and the null-groups of all* $y \in T$ *coincide with that of* x.

In case 3: $T = x \mathsf{T} [0, \lambda]$, *and for all* $y \in T$ *the* +*solution* t_y^+ *is 1–1 on* $[0, \lambda)$.

In case 4: *for all* $y \in T$, t_y^+ *is 1–1.*

Proof. Apply 4.4 repeatedly.

4.10. *Corollary. If* T *is an lsd system with -unicity on a countable carrier* P, *then every point of* P *is critical or feebly critical.*

Proof. 4.6, 4.9.3–4.

In a more physical terminology, 4.5 and 4.10 may be interpreted as follows. On a set of states, consider an autonomous system with non-discrete time, deterministic into both the future and the past (but without any continuity assumptions). If the set of states is finite, the system is completely immobile. If the set of states is countable, then the system admits at most a type of Brownian movement.

4.11. *Corollary. Let* T *be an lsd system with -unicity on* P. *An* $x \in P$ *is feebly critical iff there exist* $y \in P$, $\theta_n \in \mathsf{R}^1$ *with*

$$x \neq y\,, \quad 0 \neq \theta_n \to 0\,, \quad x = y \mathsf{T} \theta_n\,.$$

Proof. 4.4, 4.9.2.

4.12. *Corollary. If* T *is an ld system on* P, *then the conclusions of 4.9 hold with* T *replaced by* $x \mathsf{T} \mathsf{R}^1$ (*and* t_y^+ *by* t_y).

Proof. 4.9, 4.2.

In particular, then, a trajectory of an ld system either consists of a single critical point or contains no critical points, and either consists of feebly critical points or contains none of these; furthermore, distinct trajectories are disjoint.

4.13. Assume given an lsd system with -unicity T on P. Define a relation $\sim$ on P by letting $x \sim y$ iff $x = x \mathsf{T} \theta_n$ for some $\theta_n \to 0$ in R^+. Obviously $\sim$ is reflexive ($\theta_n \equiv 0$) and transitive and, according to 4.9.3, it is symmetric; obviously $\sim$ is compatible with T, and from $\alpha_x = +\infty$ in 4.3 it even follows that $\sim$ is strongly compatible with T. Hence one may apply construction 3.5, resulting in an lsd system T′ on $P/\sim$. It is then obvious and easily proved that the new lsd system has no feebly critical points. Furthermore, in the transition from (P, T) to $(P/\sim, \mathrm{T}')$, each feebly critical trajectory is collapsed into a single critical point and all other points of P are reproduced perfectly. In particular -unicity is preserved, and s-points of P and of $P/\sim$ coincide. Evidently a similar construction with similar conclusions may be carried out for ld systems.

In either case, the resulting object $(P/\sim, \mathrm{T}')$ will be termed the *T_0-modification* of (P, T); the reason for this terminology will become apparent in III,1.11. The process of forming the T_0-modification may be described as the action of covariant functors

$$\mathscr{D}_l \to \mathscr{D}_l \quad \text{and} \quad \mathscr{D}_{slu} \to \mathscr{D}_{slu}$$

commuting with the restriction functors of 3.12.

4.14. Next, consider the notions introduced in 3.17, and let T be an lsd system on P symmetric relative to a symmetry $f : P \to P$ (*loc. cit.*). The invariance property may be formulated in terms of the translations T_θ (cf. 1.4)

$$f\,T_\theta(x) = f(x \,\mathrm{T}\, \theta) = f(x)\,\mathrm{T}\,\theta = T_\theta f(x)$$

as a commutativity condition, $f T_\theta = T_\theta f$. It follows that f maps the set of fixed points of T_θ into itself; in fact the map is onto, since from $T_\theta f(x) = f(x)$ there follows

$$x = f f(x) = f\,T_\theta f(x) = ff\,T_\theta(x) = T_\theta(x)\,.$$

Similarly, for an ld system anti-symmetric relative to a symmetry f one has $f T_\theta = T_{-\theta} f$, and from $T_\theta f(x) = f(x)$ there follows

$$x = f f(x) = f\,T_\theta f(x) = ff\,T_{-\theta}(x) = T_{-\theta}(x)\,.$$

Now, critical points may be characterized in terms of fixed points of translations (x is a critical point iff it is a common fixed point of all T_θ with $\theta > 0$, cf. 4.6.1), and similarly for feebly critical points, etc. Hence one has

4.15. Lemma. Let T *be an lsd system on* P*, symmetric relative to a symmetry* $f : P \to P$*. Then* f *maps the set of critical points of* T *onto itself, and similarly for the set of feebly critical points, the set of points on cycles with fixed period* λ*, and the set of points on non-cyclic trajectories. The same conclusions hold for an ld system anti-symmetric relative to an* $f : P \to P$.

4.16. Symmetry principle. Let $f : P \to P$ *be a symmetry. If* T *is an lsd system on* P *symmetric relative to* f*, and if*

$$f(x) = x \,\mathrm{T}\, \lambda \tag{4}$$

for some $x \in P$, $\lambda > 0$, *then* $x = x \top 2\lambda$. *If* $\top$ *is an ld system on P anti-symmetric relative to f, and if simultaneously*

$$f(x) = x\,, \quad f(x \top \lambda) = x \top \lambda \tag{5}$$

for some $x \in P$, $\lambda > 0$, *then again* $x = x \top 2\lambda$. *Therefore, in either case, x is either critical or feebly critical or on a cycle with period* $\leqq 2\lambda$.

Proof. In the first case, from (4) follows

$$x = f f(x) = f(x \top \lambda) = f(x) \top \lambda = (x \top \lambda) \top \lambda = x \top 2\lambda\,;$$

in the second case, from (5)

$$x \top \lambda = f(x \top \lambda) = f(x) \top (-\lambda) = x \top (-\lambda)$$

and hence again $x \top 2\lambda = x$.

5. INVARIANCE

5.1. *Definition.* Let $\top$ be an lsd system on P. A subset $X \subset P$ is called *+invariant* if

$$X \top \mathsf{R}^+ \subset X\,; \tag{1}$$

−invariant sets are defined as the complements of +invariant sets; subsets simultaneously +invariant and −invariant are called *invariant.*

Some elementary properties of +invariant, etc., sets will now be presented. Since always $X \top \mathsf{R}^+ \supset X \top 0 = X$, an equivalent formulation of (1) is

$$X = X \top \mathsf{R}^+\,.$$

5.2. *Lemma. If* $\top$ *is an lsd system on P, then an* $X \subset P$ *is −invariant iff*

$$x \top \theta \in X\,, \quad \theta \in \mathsf{R}^+ \quad \text{imply} \quad x \in X\,. \tag{2}$$

Proof. Let $X = P - Y$ be −invariant, so that Y is +invariant, and assume the premises of (2). Then $x \notin X$ would imply $x \in Y$ and, from +invariance, $x \top \theta \in Y$ and $x \top \theta \notin X$, a contradiction. Hence $x \in X$, proving (2); and similarly for the converse assertion.

5.3. *Lemma. Let* $\top$ *be an ld system on P and* $X \subset P$. *Then X is −invariant iff* $X = X \top \mathsf{R}^-$; *X is invariant iff* $X = X \top \mathsf{R}^1$.

Proof. 5.3.1. Let X be −invariant, and assume $x \top \theta$ defined for an $x \in X$, $\theta \in \mathsf{R}^-$. Then (2) applied to

$$(x \top \theta) \top (-\theta) = x \in X\,, \quad -\theta \in \mathsf{R}^+\,,$$

yields $x \mathsf{T} \theta \in X$; hence $X \mathsf{T} \mathsf{R}^- \subset X$, and from $X = X \mathsf{T} 0 \subset X \mathsf{T} \mathsf{R}^-$ there then follows $X = X \mathsf{T} \mathsf{R}^-$. Conversely, let $X = X \mathsf{T} \mathsf{R}^-$, $x \mathsf{T} \theta \in X$, $\theta \in \mathsf{R}^+$; then

$$x = (x \mathsf{T} \theta) \mathsf{T} (-\theta) \in X \mathsf{T} \mathsf{R}^- = X$$

by assumption, so that from 5.2 X is $-$invariant.

5.3.2. If X is invariant, then from the preceding and (1),

$$X \mathsf{T} \mathsf{R}^+ = X = X \mathsf{T} \mathsf{R}^-,$$

implying

$$X = (X \mathsf{T} \mathsf{R}^+) \cup (X \mathsf{T} \mathsf{R}^-) = X \mathsf{T} (\mathsf{R}^+ \cup \mathsf{R}^-) = X \mathsf{T} \mathsf{R}^1;$$

conversely $X = X \mathsf{T} \mathsf{R}^1$ implies $X \supset X \mathsf{T} \mathsf{R}^+$ and $X \supset X \mathsf{T} \mathsf{R}^-$, i.e. $+$invariance and $-$invariance of X. This concludes the proof of 5.3.

An important application of the invariance concept appears in the two following otherwise straightforward assertions.

5.4. Lemma. Let T *be an lsd system on P, and Q a +invariant subset of P. Then the partialization of* T *to*

$$(Q \times \mathsf{R}^+) \cap \textit{domain}\ \mathsf{T}$$

is an lsd system T' *on Q; furthermore, bounds are preserved,*

$$\alpha'_x = \alpha_x \quad \textit{for} \quad x \in Q,$$

T' *has -unicity if* T *does, and each s-point of* T *in Q is an s-point of* T'.

5.5. Lemma. Let T *be an ld system on P, and Q an invariant subset of P. Then the partialization of* T *to*

$$(Q \times \mathsf{R}^1) \cap \textit{domain}\ \mathsf{T}$$

is an ld system T' *on Q, and bounds are preserved:*

$$\alpha'_x = \alpha_x, \quad \beta'_x = \beta_x \quad \textit{for} \quad x \in Q.$$

In either of the cases described in 5.4–5, the resulting dynamical system T' will be termed the *relativization* of T to Q. A more general case will be introduced in the following chapter.

Next there are listed further elementary properties of $+$invariant, etc., sets; the (omitted) proofs are entirely straightforward.

5.6. Lemma. Let T *be an lsd system on P. Then*

1. P and $\varnothing$ *are both invariant.*

2. A singleton is +invariant (or invariant) iff it is a critical point.

3. *A singleton is* $-$*invariant iff it is either critical or an s-point.*

4. *If* $X_i \subset P$ *are all* $+$*invariant (or all* $-$*invariant, invariant), then so are* $\bigcap X_i$ *and* $\bigcup X_i$.

5. *The complement of a* $+$*invariant* ($-$*invariant, invariant) set is* $-$*invariant* ($+$*invariant, invariant, respectively).*

6. *The least* $+$*invariant subset containing a given* $X \subset P$ *is* $X \mathsf{T} \mathsf{R}^+$.

7. *The least* $-$*invariant subset containing a given* $X \subset P$ *is*

$$\{y : (y \mathsf{T} \mathsf{R}^+) \cap X \neq \emptyset\} .$$

8. *If* T *is invariant relative to a map* $f : P \to P$, *then the set* F *of fixed points of* f *is* $+$*invariant; if* T *has* $-$*unicity, then* F *is invariant.*

Results concerning ld systems follow directly or on applying 5.3. In particular, for T an ld system on P, the least $+$invariant, $-$invariant, invariant sets containing a given $X \subset P$ are, respectively,

$$X \mathsf{T} \mathsf{R}^+ , \quad X \mathsf{T} \mathsf{R}^- , \quad X \mathsf{T} \mathsf{R}^1 .$$

From 5.6.6–7 there follow results on maximal $+$invariant, etc., sets contained within a given subset; for example for ld systems T, an $X \subset P$ contains no $+$trajectory iff

$$(P - X) \mathsf{T} \mathsf{R}^- = P .$$

Since the set of all s-points is $-$invariant (5.6.3–4), from 5.4 there follows

5.7. *Corollary. If* T *is an lsd system on* P *and* S *the set of all s-points, then the relativization of* T *to* $P - S$ *is an lsd system with the same bounds and no s-points.*

From 5.6.4–5 one has that the set-difference of a $+$invariant and a $-$invariant set is $+$invariant. It is then natural to unquire about properties of the set-difference of two $+$invariant sets; obviously this need be neither $+$invariant nor $-$invariant, and the class of all such differences contains all $+$invariant or $-$invariant sets. To study this we introduce a new concept, which will also be useful later.

5.8. *Definition.* Given an lsd system T on P, a subset $X \subset P$ will be termed *t-convex* if for $\theta \in \mathsf{R}^+$

$$x \in X , \quad x \mathsf{T} \theta \in X \quad \text{implies} \quad x \mathsf{T} \theta' \in X \quad \text{for all} \quad \theta' \in [0, \theta] .$$

5.9. *Lemma. Let* T *be an lsd system on* P. *The set-difference of two* $+$*invariant sets is t-convex; in particular each* $+$*invariant or* $-$*invariant set is t-convex. If* X_i *are t-convex then so is* $\bigcap X_i$. *If* T *has* $-$*unicity, then a singleton is t-convex iff it is neither feebly critical nor on a cycle.*

Proof. Let $X = X_1 - X_2$ with $+$invariant X_k, take any x, θ, θ' with

$$x \in X , \quad x \mathsf{T} \theta \in X , \quad 0 \leqq \theta' \leqq \theta ,$$

and consider whether $x \top \theta' \in X$. Since X_1 is +invariant, $x \in X \subset X_1$ and $\theta' \in \mathsf{R}^+$ imply $x \top \theta' \in X_1$. Since $P - X_2$ is −invariant, $\theta - \theta' \in \mathsf{R}^+$ and

$$(x \top \theta') \top (\theta - \theta') = x \top \theta \in X \subset P - X_2$$

imply $x \top \theta' \in P - X_2$; thus indeed $x \top \theta' \in X_1 \cap (P - X_2) = X$ as was to be proved. The remaining assertions are proved easily.

5.10. *Proposition. Let* $\top$ *be an lsd system on* P, $X \subset P$. *The least t-convex set containing* X *is* $Y = Y_1 \cap Y_2$ *with*

$$Y_1 = X \top \mathsf{R}^+, \quad Y_2 = \{y : (y \top \mathsf{R}^+) \cap X \neq \emptyset\}.$$

Proof. From 5.6, Y_1 is +invariant and Y_2 −invariant, and both contain X; then from 5.9

$$Y = Y_1 \cap Y_2 = Y_1 - (P - Y_2)$$

is t-convex (and contains X). Hence it only remains to prove that $Y \subset Z$ for any t-convex $Z \supset X$. Take any $y \in Y$; then

$$y = x \top \theta \quad \text{for some} \quad x \in X,\ \theta \in \mathsf{R}^+,$$

$$y \top \theta' \in X \quad \text{for some} \quad \theta' \in \mathsf{R}^+.$$

Therefore

$$x \in X \subset Z, \quad x \top (\theta + \theta') = y \top \theta' \in X \subset Z, \quad 0 \leqq \theta \leqq \theta + \theta';$$

from t-convexity of Z there then follows $y = x \top \theta \in Z$. This proves $Y \subset Z$ and concludes the proof.

5.11. *Corollary. X is t-convex iff it is the set-difference of two +invariant (or of two −invariant) sets.*

Proof. 5.9; 5.10 with $X = Y_1 - (P - Y_2) = Y_2 - (P - Y_1)$.

As usual, results concerning ld systems may be read off directly; and in 5.10, one has $Y_2 = X \top \mathsf{R}^-$.

In Chapter V we shall need several results which it seems appropriate to include in the present section; the question treated may be formulated as a special case of the extension problem for morphisms in dynamical categories. It will be convenient first to introduce the following definition.

5.12. *Definition.* If $\top$ is an lsd system on P and $X \subset P$, then the set $X \top \mathsf{R}^+$ is said to be *semi-generated* by X. Similarly, if $\top$ is an ld system on $P \supset X$, then $X \top \mathsf{R}^1$ is said to be *generated* by X.

A descriptive characterization of this concept appears in 5.6.6. The following lemma will also be useful later.

5.13. Lemma. If T *is an lsd system on* P *and* $Q \subset P$ *semi-generates* P*, then* Q *contains all the s-points of* (P, T).

Proof. By assumption, $P = Q\,\mathsf{T}\,\mathsf{R}^+$, so that

$$P\,\mathsf{T}\,(0, +\infty) = (Q\,\mathsf{T}\,\mathsf{R}^+)\,\mathsf{T}\,(0, +\infty) = Q\,\mathsf{T}\,(0, +\infty)\,;$$

from 2.14, the set of s-points of T is the set

$$\begin{aligned} P - (P\,\mathsf{T}\,(0, +\infty)) &= (Q\,\mathsf{T}\,\mathsf{R}^+) - (Q\,\mathsf{T}\,(0, +\infty)) \\ &= (Q \cup (Q\,\mathsf{T}\,(0, +\infty))) - (Q\,\mathsf{T}\,(0, +\infty)) \subset Q \end{aligned}$$

as asserted.

5.14. Proposition. Let there be given objects $(P, \mathsf{T}) \subset (P_0, \mathsf{T}_0)$ *in* $\mathscr{D}_{slu}$, (P', T') *in* $\mathscr{D}_{sg}$, *and a morphism* $f : (P, \mathsf{T}) \to (P', \mathsf{T}') \in \mathscr{D}_{sl}$; *and assume that* P *is t-convex in* P_0 *and that each s-point of* (P, T) *remains an s-point in* (P_0, T_0). *Then there exists in* $\mathscr{D}_{sl}$ *a unique extension* g *of* f, *defined on the subset of* P_0 *semi-generated by* P.

For this notation, refer to 3.9. The diagram appropriate to the proposition is

$$\begin{array}{ccc} (P, \mathsf{T}) & \subset & (P_0, \mathsf{T}_0) \\ & {}_{f}\searrow \quad \swarrow_{g} & \\ & (P', \mathsf{T}') & \end{array}$$

Proof. 5.14.1. The first step consists of showing that, if $x_1\,\mathsf{T}_0\,\theta_1 = x_2\,\mathsf{T}_0\,\theta_2$ with $x_1, x_2 \in P$, then $f(x_1)\,\mathsf{T}'\,\theta_1 = f(x_2)\,\mathsf{T}'\,\theta_2$. Assume for example that $\theta_1 \geqq \theta_2$; from 2.6.1,

$$x_1 \in P\,, \quad x_2 = x_1\,\mathsf{T}_0\,(\theta_1 - \theta_2) \in P\,.$$

Now prove that $x_1\,\mathsf{T}\,(\theta_1 - \theta_2)$ is defined. Assume the contrary, i.e. that $\alpha_{x_1} < \theta_1 - \theta_2$; from t-convexity, $x_1\,\mathsf{T}_0\,\alpha_{x_1}$ belongs to P, and easily $x_1\,\mathsf{T}_0\,\alpha_{x_1}$ is an s-point of (P, T) but obviously not of (P_0, T_0), a contradiction. Hence

$$x_2 = x_1\,\mathsf{T}\,(\theta_1 - \theta_2)$$

and therefore

$$f(x_2) = f(x_1)\,\mathsf{T}'\,(\theta_1 - \theta_2)\,, \quad f(x_2)\,\mathsf{T}'\,\theta_2 = f(x_1)\,\mathsf{T}'\,\theta_1$$

as asserted.

5.14.2. Set $P_1 = P\,\mathsf{T}_0\,\mathsf{R}^+$, the subset of P_0 semi-generated by P. From 5.6.1–6, P_1 is $+$invariant in P, so that using 5.4, T_0 relativizes to an lsd system T_1 on P_1 with

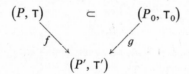

$\mathsf{T} \subset \mathsf{T}_1 \subset \mathsf{T}_0$. To any $x \in P_1$, there then exist $y \in P$, $\theta \in \mathsf{R}^+$ with $x = y\,\mathsf{T}_0\,\theta$, and one may define (since T' is global)

$$g(x) = f(y)\,\mathsf{T}'\,\theta\,.$$

From 5.15 it follows that $g : P_1 \to P'$ is indeed a map; obviously it is a morphism in $\mathscr{D}_{sl}$ with $g \mid P = f \mid P$, and evidently g is uniquely determined by these conditions. This concludes the proof of 5.14.

5.15. Corollary. To an lsd system (P, T) with $-$unicity assume there exists a gsd system (P_0, T_0) with $-$unicity again, and such that $(P, \mathsf{T}) \subset (P_0, \mathsf{T}_0)$, P is $-$invariant in P_0 and semi-generates P_0, and each s-point of P is an s-point of P_0. Then (P_0, T_0) is determined uniquely up to isomorphism in $\mathscr{D}_{sg}$.

5.16. Corollary. Assume given $(P, \mathsf{T}) \subset (P_0, \mathsf{T}_0)$ in $\mathscr{D}_l$, (P', T') in $\mathscr{D}_g$, $f : (P, \mathsf{T}) \to \to (P', \mathsf{T}')$ in $\mathscr{D}_l$. If P is t-convex in P_0, then in $\mathscr{D}_l$ there is a unique extension g of f, defined on the subset of P_0 generated by P.

5.17. Corollary. To an ld system (P, T) let there exist a gd system $(P_0, \mathsf{T}_0) \supset \supset (P, \mathsf{T})$ such that P is t-convex in P_0 and generates P_0; then (P_0, T_0) is determined uniquely up to isomorphism in $\mathscr{D}_g$.

5.18. Corollary. Assume given (P, T) in $\mathscr{D}_{sg}$, (P_0, T_0) and (P', T') in $\mathscr{D}_g$, $f : (P, \mathsf{T}) \to \to (P', \mathsf{T}')$ in $\mathscr{D}_{sg}$, and let $\mathsf{T} \subset \mathsf{T}_0$. Then in $\mathscr{D}_g$ there is a unique extension g of f, defined on the subset of P_0 generated by P.

(This is not a consequence of 5.14; however, 5.14.1 may be reproved in the present case, using the global property of T).

5.19. Corollary. To a gsd system (P, T) let there exist a gd system $(P_0, \mathsf{T}_0) \supset \supset (P, \mathsf{T})$ such that P generates P_0. Then (P_0, T_0) is determined uniquely up to isomorphism in $\mathscr{D}_g$.

5.20. Assume given an lsd system T on a set P. Denote by (P^2, T^2) the direct product of two specimens of (P, T), in the sense of 3.2; thus T^2 is an lsd system on $P^2 = P \times P$ defined by

$$(x_1, x_2)\,\mathsf{T}^2\,\theta = (x_1\,\mathsf{T}\,\theta, x_2\,\mathsf{T}\,\theta)$$

for $\theta \in \mathsf{R}^+$ whenever both $x_k\,\mathsf{T}\,\theta$ are defined. Also, let D be the diagonal of P^2,

$$D = \{(x, x) : x \in P\}\,.$$

Since, obviously, $x_1 = x_2$ implies $x_1\,\mathsf{T}\,\theta = x_2\,\mathsf{T}\,\theta$ in (P, T), one has from 5.1 that D is $+$invariant relative to T^2. Lemma 5.2 then yields the following interesting characterization.

5.21. Lemma. An lsd system T on P has $-$unicity iff the diagonal in P^2 is invariant relative to T^2.

6. SECTIONS

First consider the relation between, points on the carrier of a dynamical system, of being on the same trajectory; more precisely, let us introduce the following.

6.1. Definition. Let T be an lsd system on P. Then points $x, y \in P$ will be termed *t-equivalent*, and this in denoted by $x \perp y$, if either

$$x \in y \mathsf{T} \mathsf{R}^+ \quad \text{or} \quad y \in x \mathsf{T} \mathsf{R}^+ .$$

6.2. Lemma. If T is an lsd system with $-$unicity on P, then t-equivalence is an equivalence relation on P compatible with T, and the equivalence class containing a given $x \in P$ is the set

$$\{u : x \in u \mathsf{T} \mathsf{R}^+\} \cup (x \mathsf{T} \mathsf{R}^+) .$$

Proof. All the assertions are immediate except for transitivity of $\perp$. Thus, assume $x \perp y \perp z$; there are four cases:

6.2.1. $x = y \mathsf{T} \theta$, $y = z \mathsf{T} \theta'$, implying

$$x = (z \mathsf{T} \theta') \mathsf{T} \theta = z \mathsf{T} (\theta' + \theta) ,$$

6.2.2. $x = y \mathsf{T} \theta$, $z = y \mathsf{T} \theta'$, yielding, for $\theta' \geqq \theta$,
$$z = y \mathsf{T} (\theta + \theta' - \theta) = (y \mathsf{T} \theta) \mathsf{T} (\theta' - \theta) = x \mathsf{T} (\theta' - \theta).$$

6.2.3. $y = x \mathsf{T} \theta$, $y = z \mathsf{T} \theta'$; from 2.6.1 for $\theta' \geqq \theta$,
$$x = z \mathsf{T} (\theta' - \theta).$$

6.2.4. $y = x \mathsf{T} \theta$, $z = y \mathsf{T} \theta'$, implying
$$z = x \mathsf{T} (\theta + \theta').$$

In all cases, then, $x \perp z$ as was to be proved. This concludes the proof.

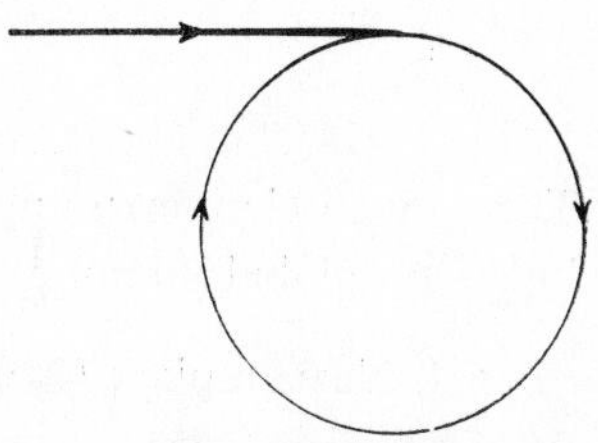

Fig. 2. *t*-equivalence on a gsd system without -unicity.

Observe that 6.2.3 is the only case in which $-$ unicity was exploited. The *t*-equivalence relation may be an equivalence relation even if T does not have $-$unicity. Thus, consider the gsd system T suggested by Fig. 2; since the carrier is a single $+$trajectory, all points are *t*-equivalent, but T does not have $-$unicity. However, in the gsd system on the triode, example 2.8, obviously *t*-equivalence is not an equivalence relation.

6.3. Corollary. Let T be an ld system on P. Then $x \perp y$ iff $x = y \mathsf{T} \theta$ for some $\theta \in \mathsf{R}^1$, $\perp$ is an equivalence relation compatible with T, and trajectories $x \mathsf{T} \mathsf{R}^1$ are the equivalence classes modulo $\perp$.

Proof. 6.2, 4.2.

One may relate 6.2 to 3.6: the factor system (P, T) modulo $\perp$ is trivial, and $\perp$ is the maximal relation compatible with T which has this property.

6.4. One may now conveniently introduce the concept of a section (local section in Niemyckij and Stepanov, 1949; cross-section in Hu, 1959); this will receive further attention in Chapters IV and VI, and assume primary importance in Chapter VII. The underlying idea of this definition may be indicated, at least vaguely, on assuming $-$unicity, as follows: a subset $Q \subset P$ is a section of T if it is a section to the relation $\perp$, relativized to some $Q \mathsf{T} [0, \varepsilon]$. (For the notion of a section Q to an equivalence relation $\sim$ on P on taking the isolated topology, refer to Bourbaki, 1951: $Q \subset P$ is a section to $\sim$ if, to any $x \in P$, there is a unique $y \in Q$ with $x \sim y$; hence Q is a more or less adequate model of $P/\sim$.)

6.5. Definition. Let T be an lsd system on P; a subset $Q \subset P$ is called a *section* if there is an $\varepsilon > 0$ such that

$$(Q \mathsf{T} \theta) \cap (Q \mathsf{T} \theta') = \emptyset \quad \text{for} \quad 0 \leqq \theta < \theta' \leqq \varepsilon . \tag{1}$$

Any such ε will be termed a *t-extent* of Q; if these ε's may be taken arbitrarily large, Q will be said to have t-extent ∞. If T is an *ld* system and Q a section of t-extent ∞ generating P, then Q will be termed a *global section.*

Examples. 6.6.1. Consider the gd system T of 1.3.3; then the y-axis, i.e. the subset $(0) \times \mathsf{R}^1$, is a global section.

6.6.2. Take the gd system defined on the complex plane R^2 by

$$\frac{dz}{d\theta} = z . \tag{2}$$

Then any circle containing the origin in its inner domain is a section of t-extent ∞ but not a global section.

6.6.3. Next, replace (2) by

$$\frac{dz}{d\theta} = iz$$

Then any open-ray issuing from the origin, i.e. the set $\{z : \operatorname{Arg} z = \alpha\}$, $\alpha \in \mathsf{R}^1$, is a section of t-extent ε for $0 \leqq \varepsilon < 2\pi$, but not for $\varepsilon \geqq 2\pi$.

6.7. Lemma. Let T be an lsd system with $-$unicity on P; then a subset $Q \subset P$ is a section of t-extent $\varepsilon > 0$ iff either of the following conditions are satisfied:

1. For any $x \in P$, $\theta \in \mathsf{R}^+$, there is at most one $\theta' \in [\theta, \theta + \varepsilon]$ with $x \mathsf{T} \theta' \in Q$.

2. $Q \cap (Q \mathsf{T} \theta) = \emptyset$ for $0 < \theta \leqq \varepsilon$.

If, furthermore, $I = [0, \varepsilon] \subset \mathsf{R}^+$ has the property that $Q \times I \subset$ domain T, there is a further equivalent condition:

3. The lsd system T *partializes to a map* $\mathsf{T}_p = \mathsf{T} \,|\, (Q \times I)$, *1–1 onto,*

$$\mathsf{T}_p : Q \times I \to Q\,\mathsf{T}\,I\,.$$

Similar results hold for t-extent ∞, *replacing* I *by* R^+.

Proof. The implications $(1) \Rightarrow 1 \Rightarrow 2 \Rightarrow (1)$ follow easily, using –unicity. As concerns 3, assume $Q \times I \subset$ *domain* T; then, in any case, the indicated partialization T_p is indeed a map onto $Q\,\mathsf{T}\,I$, so that the question is only as to whether T_p is 1–1.

If T_p is 1–1, then $\theta \neq \theta'$ in I implies

$$\begin{aligned}(Q\,\mathsf{T}\,\theta) \cap (Q\,\mathsf{T}\,\theta') &= \mathsf{T}_p(Q \times (\theta)) \cap \mathsf{T}_p(Q \times (\theta')) \\ &= \mathsf{T}_p((Q \times (\theta)) \cap (Q \times (\theta'))) = \emptyset\end{aligned}$$

since $Q \times (\theta)$ and $Q \times (\theta')$ are disjoint; hence (1). For the converse implication, assume (1) and

$$x\,\mathsf{T}_p\,\theta = x'\,\mathsf{T}_p\,\theta' \quad \text{with} \quad x, x' \in Q \quad \text{and} \quad \theta, \theta' \in I\,;$$

then $\theta = \theta'$ from (1), whereupon $x = x'$ from –unicity; thus T_p is indeed 1–1. This completes the proof of 6.7.

Condition 6.7.1 may be formulated, less accurately but perhaps more vividly, as requiring that any trajectory arc of *t*-length ε intersect Q at most once. For ld systems, occasionally another formulation of 6.7 is more useful:

6.8. Lemma. Let T *be an ld system on P; then a subset* $Q \subset P$ *is a section of t-extent* $\varepsilon > 0$ *iff either of the following conditions is satisfied.*

1. $(Q\,\mathsf{T}\,\theta) \cap (Q\,\mathsf{T}\,\theta') = \emptyset$ for $-\tfrac{1}{2}\varepsilon \leqq \theta < \theta' \leqq \tfrac{1}{2}\varepsilon$.

2. $Q \cap (Q\,\mathsf{T}\,\theta) = \emptyset$ for $0 < |\theta| \leqq \varepsilon$.

If $I = [-\tfrac{1}{2}\varepsilon, \tfrac{1}{2}\varepsilon]$ *has* $Q \times I \subset$ *domain* T, *then a third equivalent condition is*

3. The ld system T *partializes to a map* $\mathsf{T}_p = \mathsf{T} \,|\, (Q \times I)$, *1–1 onto,*

$$\mathsf{T}_p : Q \times I \to Q\,\mathsf{T}\,I\,.$$

Similar results hold for t-extent ∞, *replacing* I *by* R^1.

Proof. Analogous to 6.7; notice, however, that the upper bound in 2 is ε and not $\tfrac{1}{2}\varepsilon$ as might have been expected from a formal analogy.

Next there are presented some elementary properties of sections; the proofs are quite straightforward.

6.9. Lemma. Let T *be an lsd system on P.*

1. If Q *is a section of t-extent* $\varepsilon > 0$, *for any* $Q' \subset Q$ *and* $0 < \varepsilon' \leqq \varepsilon$, Q' *is a section of t-extent* ε'.

2. If Q *is a section of t-extent* $\varepsilon > 0$ *and* $0 \leqq \theta < \varepsilon$, *then* $Q\,\mathsf{T}\,\theta$ *is a section of t-extent* $\varepsilon - \theta$.

3. *$\emptyset$ is a section of t-extent ∞; P, if non-void, is not a section.*
4. *The set of s-points is a section of t-extent ∞.*
5. *A singleton is a section iff it is neither critical nor feebly critical.*
6. *Each t-extent of a section Q is a lower bound to periods of cycles interesting Q.*
7. *If $f:(P, \mathsf{T}) \to (P', \mathsf{T}')$ is a morphism in $\mathscr{D}_{sl}$ and Q' is a section in (P', T') with t-extent $\varepsilon' > 0$, then $f^{-1}(Q')$ is a section in (P, T) with t-extent ε'.*

In particular, from 5 and 1, a section contains no critical nor feebly critical points.

6.10. Example. The set-union of two sections may be, but often is not, a section. Thus, in example 6.6.2, the union of two distinct circles (not necessarily concentric, both containing the origin in their interior domain) is a section iff the circles do not intersect. (Further results will be given in III,1.15 and VI,1.8.)

Finally, we shall notice the behaviour of sections under the operations of forming direct products and factor systems (cf. Section 3).

6.11. Lemma. Let (P_i, T_i) be lsd systems for $i \in I$, an index set, and assume that their direct product (P, T) exists. If $Q_i \subset P_i$ and at least one Q_i is a section of t-extent $\varepsilon > 0$ in (P_i, T_i), then $Q = \Pi Q_i$ is a section of t-extent ε in (P, T). A similar assertion holds for ld systems.

Proof. 6.9.7 and 6.9.1, using the projection $p_i : P \to P_i$.

6.12. Proposition. A gd system T on P possesses a global section iff (P, T) is the direct product of a trivial gd system with the natural gd system on R^1.

6.12.1. Proof. First, let (P, T) be the direct product of a (P_0, T_0) with T_0 trivial and $(\mathsf{R}^1, +)$, and set $Q = P_0 \times (0)$. Then

$$Q \,\mathsf{T}\, \mathsf{R}^1 = \bigcup_{\theta \in \mathsf{R}^1} ((P_0 \,\mathsf{T}_0\, \theta) \times (0 + \theta)) = \bigcup_{\theta \in \mathsf{R}^1} (P_0 \times (\theta)) = P_0 \times \mathsf{R}^1 = P\,,$$

so that Q generates P. Since the singleton $(0) \subset \mathsf{R}^1$ is a section of t-extent ∞, Q is a section of t-extent ∞ (cf. 6.11). By definition then Q is a global section to (P, T).

6.12.2. Conversely, let (P, T) have a global section Q. Since T is global, one may apply 6.8.3; under the present assumptions, it results that

$$\mathsf{T} : Q \times \mathsf{R}^1 \to Q \,\mathsf{T}\, \mathsf{R}^1 = P\,, \quad 1-1 \text{ onto}\,.$$

Hence T may be interpreted as an isomorphism in $\mathscr{D}_g$ between P and the direct product of Q endowed with the trivial gd system T_0, and of $(\mathsf{R}^1, +)$. This completes the proof of 6.12.

6.13. Lemma. Let (P, T) be an lsd system, $\sim$ an equivalence relation compatible with T, (P', T') the factor system of (P, T) modulo $\sim$. Then, if Q' is a section in (P', T') with t-extent $\varepsilon' > 0$, $e^{-1}(Q')$ is a section in (P, T) with t-extent ε'. A similar assertion obtains for ld systems.

Proof. 6.9.7.

CHAPTER III

Inherent Topology

In this chapter the study of abstract dynamical systems initiated in Chapter II is continued. The first section develops an auxiliary tool, the topology of the carrier space defined immanently by the local dynamical system itself. Its main advantage is that it translates dynamical properties into topological and these latter are, of course, much better developed. The second section contains two of the local determinacy theorems; these give conditions under which the dynamical system as a whole is characterized by its local behaviour. Although the present theorems are sufficient to answer several of the problems raised in Chapter II (cf. corollaries 2.13–20), many further questions remain open, and most of these are apparently even more interesting; see 2.21.

1. INHERENT TOPOLOGY

Throughout this section we adopt the convention that there is given an ld system T on an abstract set P, in the sense of II,1.1. It will now be shown that T induces a topology on P, having interesting connections with T.

1.1. Definition. A subset $X \subset P$ will be termed *t-open* if $x \in X$ implies $x\,\mathsf{T}\,(-\varepsilon, \varepsilon) \subset X$ for some $\varepsilon > 0$. The system of t-open sets thus obtained satisfies the open-set axioms, and hence defines a topology on P; it will be termed the t-topology or *inherent* topology of T.

1.2. Qualifiers such as t-closed, t-convergent, etc., will be used with the obvious meaning (however, recall that the similarly formed non-topological terms t-convexity, t-equivalence and t-extent have been defined previously in Chapter II, and are not connected with the present t-topology).

It is easily seen that the system

$$\{x\,\mathsf{T}\,(-\varepsilon, \varepsilon) \mid \varepsilon > 0\}$$

constitutes a base at x for the t-open sets. Hence, t-convergence of generalized sequences in P may be described thus: $x_i \to x$ in the t-topology of P iff, for large i, $x_i = x\,\mathsf{T}\,\theta_i$ with $\theta_i \to 0$ in R^1. Obviously this also yields a characterization of t-closed sets. It also follows that P is t-locally quasi-compact and t-locally pathwise connected.

Obviously, a subset $X \subset P$ is invariant iff it is t-closed-open; in particular, this obtains for trajectories. Hence P is t-connected iff it consists of a single trajectory (or if $P = \emptyset$), and if P is t-quasi-compact it consists of a finite set of trajectories.

As an elementary exercise, it can be shown that the t-frontier of a t-convex set $X \supset P$ consists of precisely all points $x \mathsf{T} \theta'$ such that $x \in X$ and either

$$\theta' = \sup \{\theta : x \mathsf{T} \theta \in X\} < \alpha_x \quad \text{or} \quad \theta' = \inf \{\theta : x \mathsf{T} \theta \in X\} > \beta_x .$$

In particular, if $Q \subset P$ is $+$invariant, then the set of s-points of the relativization of T to Q (cf. II,5.4) may be described as the intersection of Q with its t-frontier in P.

1.3. Lemma. domain T *is t-open, and the map*

$$\mathsf{T} : \textit{domain}\ \mathsf{T} \to P$$

is t-continuous.

For definiteness, the topologies employed in this assertion are as follows: inherent for P, natural for R^1, the product topology for $P \times \mathsf{R}^1$, the subspace topology for *domain* $\mathsf{T} \subset P \times \mathsf{R}^1$.

Proof of 1.3. It suffices to show that the inverse image under T of any t-open $G \subset P$ is open in $P \times \mathsf{R}^1$ (for the first assertion merely take $G = P$) and to this end one need only consider $G = x \mathsf{T} (-\varepsilon, \varepsilon)$ basic t-open. Thus, assume that $(x', \theta') \in P \times \mathsf{R}^1$ T-maps into the set $x \mathsf{T} (-\varepsilon, \varepsilon)$, i.e. let

$$x' \mathsf{T} \theta' = x \mathsf{T} \theta , \quad |\theta| < \varepsilon .$$

Set $\delta = \frac{1}{2}(\varepsilon - |\theta|)$ and consider, in $P \times \mathsf{R}^1$, the set

$$(x' \mathsf{T} (-\delta, \delta)) \times (\theta' - \delta, \theta' + \delta) .$$

Evidently, this set is open in $P \times \mathsf{R}^1$, and its T-image is contained within

$$(x' \mathsf{T} (-\delta, \delta)) \mathsf{T} (\theta' - \delta, \theta' + \delta) = (x' \mathsf{T} \theta') \mathsf{T} (-2\delta, 2\delta) =$$
$$= (x \mathsf{T} \theta) \mathsf{T} (|\theta| - \varepsilon, -|\theta| + \varepsilon) = x \mathsf{T} (\theta + |\theta| - \varepsilon, \theta - |\theta| + \varepsilon) \subset x \mathsf{T} (-\varepsilon, \varepsilon) .$$

This concludes the proof of 1.3.

1.4. Theorem. The inherent topology of an ld system T *on* P *is the finest among topologies on* P *rendering all solutions* $t_x : (\beta_x, \alpha_x) \to P$ *continuous.*

Proof. Each solution t_x is essentially a partialization of the map T (cf. II,1.4), so that, according to 1.3, all t_x are indeed t-continuous. Conversely, assume that all t_x are continuous in a topology τ on P; take any $G \subset P$ open under τ, and any $x \in G$. Then $t_x(0) = x \in G$, so that $x \mathsf{T} (-\varepsilon, \varepsilon) = t_x(-\varepsilon, \varepsilon) \subset G$ for small $\varepsilon > 0$, proving that G is also open in the inherent topology of T. Thus indeed the inherent topology is finer than τ.

1.5. Corollary. The inherent topology is the finest among topologies on P rendering continuous the map T: *domain* $\mathsf{T} \to P$.

Proof. 1.3, 1.4.

1.6. Corollary. The escape times α_x, β_x *of II,1.1, considered as maps* $P \to \mathsf{R}^\#$, *are t-continuous.*

Proof. 1.4, II,1.5.

1.7. Proposition. The inherent topology of an ld system is pseudo-metrisable.

Proof. 1.7.1. Let T be the ld system on P. For $x, y \in P$ define

$$\sigma(x, y) = \inf \{|\theta| : x = y \mathsf{T} \theta\}.$$

For definiteness, if $x = y \mathsf{T} \theta$ for no $\theta \in \mathsf{R}^1$, then the indicated set is empty, and its g.l.b. is $+\infty$. Obviously $\sigma(x, x) = 0 \leqq \sigma(x, y) \leqq +\infty$. To verify symmetry $\sigma(x, y) = \sigma(y, x)$, merely observe that $x = y \mathsf{T} \theta$ implies

$$y = (y \mathsf{T} \theta) \mathsf{T} (-\theta) = x \mathsf{T} (-\theta).$$

In proving the triangle inequality

$$\sigma(x, y) + \sigma(y, z) \geqq \sigma(x, z) \tag{1}$$

one need only consider the case that both terms on the left are finite. For each positive integer n find θ, θ' in R^1 (these depend on n, of course) with

$$x = y \mathsf{T} \theta, \quad |\theta| < \sigma(x, y) + \frac{1}{n},$$

$$y = z \mathsf{T} \theta', \quad |\theta'| < \sigma(y, z) + \frac{1}{n};$$

then

$$x = y \mathsf{T} \theta = (z \mathsf{T} \theta') \mathsf{T} \theta = z \mathsf{T} (\theta' + \theta),$$

implying

$$\sigma(x, z) \leqq |\theta' + \theta| \leqq |\theta'| + |\theta| < \sigma(x, y) + \sigma(y, z) + \frac{2}{n}.$$

To prove (1) it now suffices to take $n \to +\infty$.

1.7.2. The next step is to set

$$\varrho = \frac{\sigma}{1 + \sigma} \tag{2}$$

(with the usual convention that $\varrho = 1$ if $\sigma = +\infty$). Obviously ϱ is then a pseudo-metric function on P.

1.7.3. It remains to show that ϱ induces precisely the inherent topology. Let $x_i \to x$ in the t-topology, i.e. $x_i = x \top \theta_i$ with $\theta_i \to 0$ in R^1; then

$$\varrho(x_i, x) \leqq \sigma(x_i, x) \leqq |\theta_i| \to 0 .$$

Conversely, if $\varrho(x_n, x) \to 0$ then also $\sigma(x_n, x) \to 0$; hence $\sigma(x_n, x) < +\infty$ for large n, and there exist $\theta_n \in \mathsf{R}^1$ with

$$x_n = x \top \theta_n , \quad |\theta_n| < \sigma(x_n, x) + \frac{1}{n} ;$$

but then $\theta_n \to 0$, i.e. the x_n indeed t-converge to x. This completes the proof of 1.7.

Observe that the pseudo-metric function ϱ constructed in 1.7 is invariant in the that always

$$\varrho(x, y) = \varrho(x \top \theta, y \top \theta) ,$$

if the latter term is defined. Next there are exhibited two corollaries to 1.7.

1.8. Corollary. Each ld system $\top$ *on a set* P *induces an inherent uniform structure on* P*, compatible with the inherent topology. Subsets of* $P \times P$ *of the form*

$$V_\varepsilon = \{(x, y) : x = y \top \theta, \ |\theta| < \varepsilon\}$$

constitute for $\varepsilon > 0$ *a base of the filter of vicinities; these are invariant in the sense that* $(x, y) \in V_\varepsilon$ *implies* $(x \top \theta, y \top \theta) \in V_\varepsilon$*. Also, the map* $\top$ *: domain* $\top \to P$ *is uniformly continuous, so that the solutions* t_x *are equi-uniformly continuous. The inherent uniform structure is the finest among uniform structures on* P *rendering all solutions* t_x *continuous.*

1.9. Corollary. If the inherent topology of $\top$ *is* T_0*, then it is metrisable; with the distance function* ϱ *of 1.7,* P *is a complete locally compact metric space.*

Proof. Under these assumptions, the pseudo-metric ϱ induces a T_0 and hence metrisable topology, and ϱ turns out to be a distance function. To prove completeness, assume that a sequence $x_n \in P$ has $\varrho(x_n, x_m) \to 0$ as $n, m \to \infty$. Then $\varrho(x_n, x_m) < 1$ for large n, m, and there exist $\theta_{nm} \in \mathsf{R}^1$ with

$$x_n = x_m \top \theta_{nm} , \quad \theta_{nm} \to 0 .$$

Fix m, and select a subsequence of integers $n_k \to \infty$ having $\theta_{n_k m}$ convergent in R^1; then x_{n_k}, a subsequence of the x_n, t-converges; it then follows from $\varrho(x_n, x_m) \to 0$ that x_n itself converges. Finally, as noted in 1.2, P is t-locally quasi-compact; since in the present case P is metrisable, it is locally compact. This concludes the proof of 1.9.

The last result seems to suggest that those systems which induce a T_0 inherent topology may prove interesting. This is indeed the case, and will now be examined.

1.10. Proposition. For an ld system T *on P, the following conditions are equivalent:*

1. *There are no feebly critical points in P.*
2. *There is a* T_0 *topology for P rendering continuous all solutions* t_x, $x \in P$.
3. *The inherent topology of* T *is* T_0.
4. *The inherent topology of* T *is metrisable.*

Proof. $3 \Leftrightarrow 4$ from 1.9, $2 \Leftrightarrow 3$ from 1.4 (a topology finer than a T_0 topology is itself T_0). From 1.7 it follows that the inherent topology of T is T_0 iff always $\varrho(x, y) = 0$ implies $x = y$, i.e. iff there do not exist $x \neq y$ in P with $x = y \mathrm{T} \theta_i$ and $\theta_i \to 0$ in R^1. According to II,4.11, this latter condition is equivalent with absence of feebly critical points in P. This proves $1 \Leftrightarrow 3$, and thus concludes the proof of 1.10.

1.11. As a verificatory example, in II,4.7 the carrier of the gd system T is a single trajectory consisting of feebly critical points; evidently, the inherent topology is the weakest topology of all (i.e. that having only the trivial subsets open).

A direct consequence of 1.10 and II,4.10 is that every ld system on a countable carrier and with a T_0 inherent topology is trivial. Thus the "Brownian movement" of II,4.10 is, topologically speaking, rather pathological.

1.12. An ld system T on P induces the inherent topology; one may then construct its topological T_0-modification, essentially by identifying points x, y in P which have $\varrho(x, y) = 0$, i.e. $x = y \mathrm{T} \theta_n$ with $\theta_n \to 0$ in R^1 (Bourbaki, 1948). This coincides with the construction of $P/\sim$ in II,4.13.

Another instance of a topological formulation of a dynamical-theoretic property is the following.

1.13. Proposition. Let T *be an ld system on P, and* $Q \subset P$. *Then there exists an ld system* T' *on Q with* $\mathrm{T}' \subset \mathrm{T}$ *iff Q is t-open in P; in the positive case, the inherent topology of* T' *coincides with the subspace topology of Q in P.*

Proof. 1.13.1. First assume that an ld system T' as described exists. From definition II,1.1.1 directly, if $x \in Q$ then $x \mathrm{T}' \theta \in Q$ for small $|\theta|$; since $\mathrm{T}' \subset \mathrm{T}$, one has that $x \mathrm{T} \theta \in Q$ for small $|\theta|$. Hence by 1.1, Q is *t*-open.

1.13.2. Let Q be *t*-open in P. Take any $x \in Q$; for $\theta' \geqq 0$ define $x \mathrm{T}' \theta = x \mathrm{T} \theta$ if $x \mathrm{T} \theta' \in Q$ for all $\theta' \in [0, \theta]$, and similarly for $\theta' \leqq 0$; otherwise leave $x \mathrm{T}' \theta$ undefined. Obviously $\mathrm{T}' \subset \mathrm{T}$ and, easily, T' is an ld system on Q; for example the bounds α'_x, β'_x of T' are determined as

$$\alpha'_x = \sup \{\theta : x \mathrm{T} \theta' \in Q \text{ for } 0 \leqq \theta' \leqq \theta\}, \text{ etc.}$$

Obviously the inherent topology of T' is as indicated. This concludes the proof of 1.13.

In the situation described in 1.13, T' will be said to have been obtained from T by *relativization* to Q. Dynamical concepts relating to T' in Q may then be termed *relative*; for example relative invariance, etc. Since invariant sets are t-open, the construction of II,5.5 is a special case of that in 1.13; however, here there is no question of the equality of bounds applying in 1.13.

1.14. Lemma. Let $\top$ *be an ld system on* P, *and* $Q \subset P$ *a t-open subset. Then one has the following assertions for subsets* $X \subset Q$: *relative t-convexity is equivalent with t-convexity; invariance, +invariance, −invariance imply the corresponding relative properties; a relatively invariant* X *is invariant iff it is t-closed.*

Proof. Possibly only the last assertion has a non-immediate proof. If X is invariant, it is t-closed by 1.2. Conversely, let $X \subset Q$ be t-closed and relatively invariant, and assume $x \top \theta \notin X$ for some $x \in X$, $\theta \in \mathsf{R}^+$. Set

$$\theta' = \inf \{\theta \in \mathsf{R}^+ : x \top \theta \notin X\} ; \tag{3}$$

then $0 < \theta' < +\infty$ by assumption, and $x \top \theta' \in X$ since X is t-closed. From $X \subset Q$ with Q t-open and X relatively +invariant there then follows $x \top (\theta' + \varepsilon) \in X$ for small $\varepsilon > 0$, contradicting (3). Hence $x \top \theta \in X$ and X is invariant as asserted.

Finally consider the sections introduced in the preceding chapter and, in particular, the situation described in II,6.10.

1.15. Lemma. Let $\top$ *be an ld system on* P *and* $Q = Q_1 \cup Q_2$ *a disjoint decomposition of a section* Q. *Then* Q_1, Q_2 *are separated by t-open sets.*

Proof. Let Q have t-extent $2\varepsilon > 0$. Then, for $k = 1, 2$,

$$U_k = Q_k \top (-\varepsilon_k, \varepsilon_k)$$

is a t-open neighbourhood of Q_k. To show that the U_k are disjoint, if one had

$$x_k \in Q_k\,, \quad |\theta_k| < \varepsilon\,, \quad x_1 \top \theta_1 = x_2 \top \theta_2 \in U_1 \cap U_2\,,$$

then $x_1 = x_2 \top (\theta_2 - \theta_1)$; thus either $\theta_1 = \theta_2$, whereupon $x_1 = x_2$ contradicts disjointness of the Q_k; or $0 < |\theta_2 - \theta_1| < 2\varepsilon$, and then

$$\emptyset \neq Q_1 \cap (Q_2 \top (\theta_2 - \theta_1)) \subset Q \cap (Q \top (\theta_2 - \theta_1))$$

contradicts II,6.8.2.

2. LOCAL DETERMINACY THEOREMS

The problem treated in this section may be formulated, somewhat inaccurately, as whether or not a dynamical system is completely determined by its "local behaviour". Three versions of the notion of local behaviour will appear, in the conditions of theorem 2.1 below, in 2.10 and in 2.13 (also see VI, 1.9).

2.1. First local determinacy theorem. Let $\mathsf{T}_1, \mathsf{T}_2$ *be ld systems on the same carrier* P*, locally equivalent in the following sense: to every* $x \in P$ *there is an* $\varepsilon_x > 0$ *such that*

$$x \mathsf{T}_1 \theta = x \mathsf{T}_2 \theta \quad \text{whenever} \quad |\theta| < \varepsilon_x . \tag{1}$$

Then $\mathsf{T}_1 = \mathsf{T}_2$.

Proof. This will consist of a number of simple steps; the inherent topology introduced in section 1 will appear as an auxiliary tool.

2.1.1. The t-topologies defined by the two ld systems coincide. Indeed, sets of the form $x \mathsf{T}_k(-\varepsilon, \varepsilon)$ with $x \in P$, $0 < \varepsilon < \varepsilon_x$ (and $k = 1, 2$, fixed) are t-open and constitute a base of the t-topology of T_k (cf. 1.2). From (1), for $k = 1, 2$, these open bases coincide, and hence so do the corresponding topologies. In particular then a notation such as $x_i \to x$ will be unambiguous.

2.1.2. Critical points, and also feebly critical points, of T_1 *and* T_2 *coincide.* Again this follows directly from (1) and II, 4.6. Thus one may speak of feebly critical points without specifying which one of the given ld systems is meant.

2.1.3. The null-groups (cf. II,4.4) *of feebly critical points in* $\mathsf{T}_1, \mathsf{T}_2$ *coincide.* Essentially this reduces to showing that a dense subgroup of R^1 is generated by any neighbourhood of zero. Take a feebly critical $x \in P$, let G_k denote the null-group of x in T_k, $k = 1, 2$. From II,4.9, each G_k is a non-trivial dense subgroup of R^1; from (1), there is a neighbourhood $U = (-\varepsilon_x, \varepsilon_x)$ of 0 such that

$$G_1 \cap U = G_2 \cap U .$$

In particular, one may choose a sequence $\theta_n \in \mathsf{R}^1$ with

$$\theta_n \in G_k \cap U , \quad 0 < \theta_n \to 0 .$$

Now take an arbitrary $\theta \in G_1$; denote by k_n the integral part of θ/θ_n, so that $\theta'_n = \theta - k_n\theta_k$ has $0 \leqq |\theta'_n| < \theta_n \to 0$. In particular, $\theta'_n \in U$ for large n; since all θ, θ_n are in G_1, one also has $\theta'_n \in G_1$, and hence

$$\theta'_n \in G_1 \cap U = G_2 \cap U \subset G_2 .$$

But also $\theta_n \in G_2$, so that

$$\theta = \theta'_n + k_n\theta_n \in G_2 .$$

Since $\theta \in G_1$ was arbitrary, this proves $G_1 \subset G_2$, and of course $G_2 \subset G_1$ by symmetry; and this was the assertion.

2.1.4. If x *is feebly critical, then*

$$x \mathsf{T}_1 \theta = x \mathsf{T}_2 \theta \quad \textit{for all} \quad \theta \in \mathsf{R}^1 .$$

Recall from II,4.3 that the present assumptions yield

$$\alpha_x^1 = +\infty = \alpha_x^2, \quad \beta_x^1 = -\infty = \beta_x^2$$

(α_x^k will denote the escape time within the k-th system T_k). As in 2.1.3 set $U = (-\varepsilon_k, \varepsilon_x)$, and let G be the common null-group of T_k. Take any $\theta \in \mathsf{R}^1$; from II,4.9, for each $k = 1, 2$, there is a $\theta_k \in \mathsf{R}^1$ with

$$x \,\mathsf{T}_k\, \theta = x \,\mathsf{T}_k\, \theta_k, \quad \theta - \theta_k \in G, \quad \theta_k \in U. \tag{2}$$

Then $\theta_2 - \theta_1 = (\theta - \theta_1) - (\theta - \theta_2) \in G$, so that $x = x \,\mathsf{T}_1\, (\theta_2 - \theta_1)$; using (1) and $\theta_k \in U$,

$$x \,\mathsf{T}_2\, \theta_2 = x \,\mathsf{T}_1\, \theta_2 = (x \,\mathsf{T}_1\, (\theta_2 - \theta_1)) \,\mathsf{T}_1\, \theta_1 = x \,\mathsf{T}_1\, \theta_1 .$$

With (2) this yields the asserted relation.

Having proved the assertion of the theorem at feebly critical points, next notice the other extreme, namely dynamical systems without feebly critical points (cf. 1.10).

2.1.5. If the inherent topology is T_0 then, for all $x \in P$,

$$x \,\mathsf{T}_1\, \theta = x \,\mathsf{T}_2\, \theta \quad \textit{for} \quad 0 \leqq \theta < \min(\alpha_x^1, \alpha_x^2). \tag{3}$$

To prove this, fix any $x \in P$ and set

$$\varepsilon = \sup\{\varepsilon' : x \,\mathsf{T}_1\, \theta = x \,\mathsf{T}_2\, \theta \text{ for } 0 \leqq \theta \leqq \varepsilon'\}, \tag{4}$$

so that

$$0 < \varepsilon_x \leqq \varepsilon \leqq \min(\alpha_x^1, \alpha_x^2).$$

If one had, contrary to (3), that $\varepsilon < \alpha_x^1, \alpha_x^2$, then both $x \,\mathsf{T}_k\, \varepsilon$ would be defined. Take $\theta_k \in \mathsf{R}^+$ with $\varepsilon < \theta_n \to \varepsilon$; from (4) and 2.1.1,

$$x \,\mathsf{T}_1\, \varepsilon \leftarrow x \,\mathsf{T}_1\, \theta_n = x \,\mathsf{T}_2\, \theta_n \to x \,\mathsf{T}_2\, \varepsilon .$$

By assumption and 1.9, the inherent topology is metric, so the limits are defined uniquely, and hence $x \,\mathsf{T}_1\, \varepsilon = x \,\mathsf{T}_2\, \varepsilon$. Applying (1) to this latter point,

$$\begin{aligned} x \,\mathsf{T}_1\, (\varepsilon + \theta) &= (x \,\mathsf{T}_1\, \varepsilon) \,\mathsf{T}_1\, \theta = (x \,\mathsf{T}_1\, \varepsilon) \,\mathsf{T}_2\, \theta \\ &= (x \,\mathsf{T}_2\, \varepsilon) \,\mathsf{T}_2\, \theta = x \,\mathsf{T}_2\, (\varepsilon + \theta) \end{aligned}$$

for sufficiently small $\theta \geqq 0$. This contradicts the construction of ε in (4), and thus concludes the proof of 2.1.5.

2.1.6. If the inherent topology is T_0 then $\mathsf{T}_1 = \mathsf{T}_2$. Since in 2.1.5 one may also perform a change of orientation, (3) also holds for $0 \geqq \theta \geqq \max(\beta_x^1, \beta_x^2)$. Thus it only remains to prove that $\alpha_x^1 = \alpha_x^2$ and $\beta_x^1 = \beta_x^2$. Assume for example that $\alpha_x^1 < \alpha_x^2$; thus $y = x \,\mathsf{T}_2\, \alpha_x^1$ is defined. Take $\theta_n \in \mathsf{R}^+$ with $\alpha_x^1 > \theta_n \to \alpha_x^1$; then

$$x \,\mathsf{T}_1\, \theta_n = x \,\mathsf{T}_2\, \theta_n \to x \,\mathsf{T}_2\, \alpha_x^1 = y,$$

so that from 1.6 and II,1.5,

$$\alpha_y^1 \leqq \liminf_n \alpha^1_{x\mathsf{T}_1\theta_n} = \lim_n (\alpha_x^1 - \theta_n) = 0\,.$$

However, this contradicts $\alpha_y^1 > 0$ in the definition of ld systems, II,1.1.1, and thus completes the proof of 2.1.6.

2.1.7. In any case $\mathsf{T}_1 = \mathsf{T}_2$. Let "construction" refer to II,4.13, i.e. to the construction of the T_0-modification of an ld system. From (1) it follows that both T_k define the same equivalence relation $\sim$ of the construction, and hence the same set $P' = P/\sim$. The construction then yields two ld systems T_k' on P' without feebly critical points; furthermore, obviously the T_k' are again locally equivalent in the sense of (1). From 1.10, the inherent topologies of T_k' are T_0, and then from 2.1.6, $\mathsf{T}_1' = \mathsf{T}_2'$.

Since in the construction (P', T_k') coincides with (P, T_k) except at feebly critical points, from $\mathsf{T}_1' = \mathsf{T}_2'$ it follows that in P, $x\,\mathsf{T}_1\,\theta = x\,\mathsf{T}_2\,\theta$ whenever either side is defined, with the possible exception of feebly critical points $x \in P$. However, this exceptional case is completely covered by 2.1.4. This completes the proofs of 2.1.7 and also of 2.1.

2.2. Corollary. If $\mathsf{T}_1 \subset \mathsf{T}_2$ *are ld systems on the same carrier, then* $\mathsf{T}_1 = \mathsf{T}_2$.

Proof. Apply 2.1 with $\varepsilon_x = \min(\alpha_x^1, -\beta_x^1)$. This is the result promised in II,1.8. Hence,

2.3. Corollary. A non-global ld system T *cannot be augmented to a gd system* $\mathsf{T}' \supset \mathsf{T}$ *on the same carrier.*

The following four results are direct consequences of theorem 2.1.

2.4. Corollary. The direct product T *of ld systems* T_i *(if defined, cf. II,3.2) is determined uniquely by the condition*

$$[x_i]\,\mathsf{T}\,\theta = [x_i\,\mathsf{T}_i\,\theta] \quad \textit{for small} \quad |\theta|\,.$$

2.5. Corollary. The direct sum T *of ld systems* T_i *is determined uniquely by the condition*

$$x\,\mathsf{T}\,\theta = x\,\mathsf{T}_i\,\theta \quad \textit{for small} \quad |\theta|\,,$$

with notation as in II,3.4.

2.6. Corollary. The factor system T' *of an ld system* T *is determined uniquely by the condition*

$$e(x\,\mathsf{T}\,\theta) = e(x)\,\mathsf{T}'\,\theta \quad \textit{for small} \quad |\theta|\,,$$

with notation as in II,3.6.

2.7. Corollary. In 1.13, the relativized ld system T' *on* $Q \subset P$ *is determined uniquely by the condition* $\mathsf{T}' \subset \mathsf{T}$.

2.8. Corollary. In II,2.9 the ld system T' *with given semi-dynamical restriction* T *is determined uniquely by this condition.*

Proof. Indeed, let T_1, T_2 be ld systems on P with coinciding lsd restrictions T; it suffices to verify that they are locally equivalent in the sense of (1). Take any $x \in P$, and then θ' with $\beta_x^1 < \theta' < 0$; then $y = x\,\mathsf{T}_1\,\theta'$ is defined and

$$x = y\,\mathsf{T}_1\,(-\theta') = y\,\mathsf{T}\,(-\theta') = y\,\mathsf{T}_2\,(-\theta').$$

Now take any $\theta \in \mathsf{R}^1$ with $\theta' < \theta < \alpha_x^1$; then

$$\begin{aligned} x\,\mathsf{T}_1\,\theta &= (x\,\mathsf{T}_1\,\theta')\,\mathsf{T}_1\,(\theta - \theta') = y\,\mathsf{T}\,(\theta - \theta') \\ &= y\,\mathsf{T}_2\,(\theta - \theta') = (y\,\mathsf{T}_2\,(-\theta'))\,\mathsf{T}_2\,\theta = x\,\mathsf{T}_2\,\theta\,. \end{aligned}$$

Thus indeed (1) holds with for example $\varepsilon_x = \min(\alpha_x^1, -\beta_x^1, -\beta_x^2)$.

2.9. It follows from the local determinacy theorem that an ld system is completely determined by the behaviour of each solution t_x on some neighbourhood $(-\varepsilon_x, \varepsilon_x)$ of 0. This is the parallel in abstract dynamical system theory of the assertion that a differential ld system as in II,1.3.1 is completely determined by the behaviour of its solutions in some first-order differential neighbourhood of the origin, i.e. by the differential equation itself.

As a further (and more or less direct) consequence of 2.1 one has the second local determinacy theorem below, applying to lsd systems. The examples which then follow treat the assumptions appearing there.

2.10. Second local determinacy theorem. Let T_1, T_2 *be lsd systems with* $-$*unicity on the same carrier P and with coinciding s-points. If to any* $x \in P$ *there is an* $\varepsilon_x > 0$ *such that for* $0 \leqq \theta < \varepsilon_x$*:*

$$x\,\mathsf{T}_1\,\theta = x\,\mathsf{T}_2\,\theta \tag{5}$$

and

$$y\,\mathsf{T}_1\,\theta = x \quad \textit{iff} \quad y\,\mathsf{T}_2\,\theta = x \tag{6}$$

then $\mathsf{T}_1 = \mathsf{T}_2$.

Proof. 2.10.1. Let S be the common set of s-points. According to II,5.7 and II,5.4, on $P' = P - S$ as carrier, each T_k relativizes to an lsd system T_k' with $-$unicity and no s-points. From II,2.9, each T_k' is the lsd restriction of an ld system T_k'' on P' (unique, cf. 2.8). From (5) applied only at $x \in P' = P - S$ and from the construction of II,2.9, it follows that the ld systems T_k'' satisfy the local equivalence condition (1), so that $\mathsf{T}_1'' = \mathsf{T}_2''$ by 2.1. Hence also $\mathsf{T}_1' = \mathsf{T}_2'$, i.e.

$$x\,\mathsf{T}_1\,\theta = x\,\mathsf{T}_2\,\theta \tag{7}$$

whenever either side is defined, for all non-s-points $x \in P$.

2.10.2. It remains to prove (7) at an arbitrary s-point x. Of course, (7) holds for $\theta = 0$; thus consider $\theta > 0$. Then there exists a $\theta' < \varepsilon_x$ with $0 < \theta' < \theta$. From (5), $x \mathsf{T}_1 \theta' = x \mathsf{T}_2 \theta'$ and, since this is a non-s-point, according to 2.10.1 for $x' = x \mathsf{T}_k \theta'$ there holds

$$x' \mathsf{T}_1 (\theta - \theta') = x' \mathsf{T}_2 (\theta - \theta');$$

thus

$$\begin{aligned} x \mathsf{T}_1 \theta &= (x \mathsf{T}_1 \theta') \mathsf{T}_1 (\theta - \theta') = x' \mathsf{T}_1 (\theta - \theta') \\ &= x' \mathsf{T}_2 (\theta - \theta') = (x \mathsf{T}_2 \theta') \mathsf{T}_2 (\theta - \theta') = x \mathsf{T}_2 \theta \end{aligned}$$

as required. This completes the proof of 2.10.

2.11. Corollary. If $\mathsf{T}_1 \subset \mathsf{T}_2$ are lsd systems with $-$unicity and coinciding carriers and s-points, then $\mathsf{T}_1 = \mathsf{T}_2$.

2.12. If the s-point sets do not coincide (i.e. T_1 has more s-points than T_2, cf. II,2.14) and the remaining assumptions are preserved, then in 2.11 one may well have $\mathsf{T}_1 \neq \mathsf{T}_2$; a simple instance of this situation was exhibited in II,2.11. Similarly, $\mathsf{T}_1 \neq \mathsf{T}_2$ may occur if T_2 does not have $-$unicity; for an example, modify example II,2.8 in analogy with II,2.11 by "interrupting" the lower branch of the triode at the branch-point. The following example shows that in 2.10 one cannot weaken the local equivalence condition appearing there to (5) only.

2.13. Example. Let T_1 be the direct sum (cf. II,3.4) of two specimens of the natural gd system $+$ on R^1; the carrier of T_1 may be identified with the disjoint union

$$P = (\mathsf{R}^1 \times (-1)) \cup (\mathsf{R}^1 \times (1)).$$

On the same carrier define T_2 thus (let $k = \pm 1$): if both x, $x \mathsf{T} \theta < 0$ or if both $x, x + \theta \geqq 0$, set $(x, k) \mathsf{T}_2 \theta = (x + \theta, k)$, and in the remaining cases set $(x, k) \mathsf{T}_2 \theta = (x + \theta, -k)$ (thus, vaguely speaking, the trajectories of T_2 switch at 0 from one branch of the carrier to the other). Then obviously T_1, T_2 are distinct gd systems, unilaterally locally equivalent in the sense that they satisfy (5), for example with $\varepsilon_{(x,k)} = +\infty$ for $x \geqq 0$, $\varepsilon_{(x,k)} = -x$ for $x < 0$.

2.14. The local determinacy theorems suggest further questions, some of which will now be indicated.

Define elementary dynamical systems (ed systems, a term devised *ad hoc*) similarly as ld systems, but with the group property II,1.1.3 relaxed to

$$(x \mathsf{T} \theta) \mathsf{T} \theta' = x \mathsf{T} (\theta + \theta')$$

required only if all the terms are defined; in particular it is no longer necessary to have $\alpha_{x \mathsf{T} \theta} = \alpha_x - \theta$ as in II,1.5.

Examples of ed systems may be obtained by taking an ld system T, choosing new bounds β_x', α_x' subject to

$$\beta_x < \beta_x' < 0 < \alpha_x' < \alpha_x$$

but otherwise arbitrary, and then partializing T. These two systems are then locally equivalent in the sense of (1). One may even formulate the original version of (1) as requiring the existence of a third ed system T with $\mathrm{T}_1 \supset \mathrm{T} \subset \mathrm{T}_2$.

Now, theorem 2.1 states that there is at most one ld system T locally equivalent to a given ed system T'. The question then arises of finding reasonable conditions on T' for the existence of at least one such T (can one even take $\mathrm{T} \supset \mathrm{T}'$ or $\mathrm{T} \subset \mathrm{T}'$?).

A second group of problems may be illustrated by the following. Admitting the presence of an ld system T locally equivalent to a given ed system T', under what conditions on T' is T global? The analogy with differential equation theory (the preceding is, essentially, the abstract counterpart to the extendability problem, I,1.11) seems to suggest that such problems, requiring locally formulated answers to questions on global properties, are of principal importance.

CHAPTER IV

Continuous Dynamical Systems

For the motivation of this chapter, see the Introduction and I,6. Essentially, a continuous dynamical system consists of a dynamical system and a topology, both on the same carrier set. The particular form of symbiosis required in 1.1 of these two structures appears to be both natural (cf. 1.3.2) and fruitful.

Sections 1 and 2 present the elementary properties of these systems, with attention focused on either the dynamical or the topological aspects. In particular, theorem 2.5 describes the connection between continuous ld and lsd systems, thereby paralleling II,2.9. Section 3 treats, among other topics, direct products and factorization for continuous dynamical systems. There seem to be two main differences in comparison with the theory for abstract systems. In general, the factor system of a continuous dynamical system is not compatible with the corresponding quotient topology; at least, a sufficient condition is given. This disadvantage is, possibly, offset by the provoking property exhibited in 3.13, which may be termed the dynamical system covering property.

The concluding Section 4 introduces two classical concepts, limit sets and orbital stability, but contains little more than the most immediate results; deeper theorems are delayed to later chapters.

The definition of a continuous dynamical system is topologically invariant, in the sense that the homeomorph of a continuous dynamical system is again such (for a more precise formulation, see 3.12). In particular, the homeomorph of a differential dynamical system on an n-manifold is a continuous dynamical system. This is strong evidence for considering continuous dynamical systems the most natural objects on which to study topological properties of differential dynamical systems. The question would be settled if one could show that this concept is the most economical one; in other words, if the following conjecture were proved: Each continuous ld system on an n-manifold is isomorphic (in the category $\mathscr{D}_l^T$) to some differential ld system. Only in extremely special cases has this been proved. (To forestall misunderstanding, this problem is not solved in the present book.)

1. CONTINUOUS DYNAMICAL SYSTEMS

On a set P, let there be given two structures, an ld system T and a topology τ. Then τ induces a product topology on $P \times \mathsf{R}^1$ (using the natural one for R^1), and hence a subset topology on *domain* $\mathsf{T} \subset P \times \mathsf{R}^1$; similar remarks apply to T an lsd system, with R^+ replacing R^1. The following definition then links the given dynamical system with these three topologies.

1.1. Definition. A topology τ and an ld (or lsd) system T, both on a set P, are termed *compatible*, and T is termed a *continuous ld system* on (P, τ) (*continuous lsd system*, respectively), if

1. the map $\mathsf{T} : domain\ \mathsf{T} \to P$ is continuous, and
2. *domain* T is open in $P \times \mathsf{R}^1$ (for lsd systems, in $P \times \mathsf{R}^+$).

If the topology τ is given or immaterial, the abbreviated form "continuous ld system on P" will often be used. Note that "continuous ld system" is to be taken as an indecomposable term; however, in the special case that T is global, condition 2 is fulfilled automatically, and a continuous gd system is then simply a continuous map with the gd properties of II,1.1.

A condition equivalent to 1 and 2 is the following, stated, for example, for ld systems: If $x_i \to x$ in P, $\theta_i \to \theta$ in R^1 and $x \mathsf{T} \theta$ is defined, then $x_i \mathsf{T} \theta_i \to x \mathsf{T} \theta$ (the latter implies, of course, that $x_i \mathsf{T} \theta_i$ is defined for large i). Another equally obvious reformulation will occasionally be more useful.

1.2. Lemma. Conditions 1.1.1–2 are equivalent to 1′: the inverse image under T *of any open subset of P is open in $P \times \mathsf{R}^1$ (in $P \times \mathsf{R}^+$ for lsd systems).*

Examples. 1.3.1. According to III,1.3, the inherent topology of an ld system T is compatible with T. An lsd system T is compatible with the isolated topology of its carrier iff T is trivial. An lsd system T is compatible with the weakest topology of its carrier iff T is global. Any topology is compatible with the trivial gd system.

A more interesting example, which, moreover, motivates most of this book and definition 1.1 in particular, is the following.

1.3.2. Let $G \subset \mathsf{R}^n$ be open, $f : G \to \mathsf{R}^n$ continuous, and consider the autonomous differential equation on G

$$\frac{dx}{d\theta} = f(x)\,; \tag{1}$$

concerning solutions of (1) assume only local unicity (cf. I,1). Construct the abstract ld system T associated with (1), in the natural manner described in II,1.3.1. Then 1.1.1–2 may be readily read off from I,1.9, on taking the natural subset topology for $G \subset \mathsf{R}^n$; note that 1.1.1 corresponds to continuous dependence on initial data.

Hence T is a continuous ld system on G; in particular, there are no feebly critical points in G (cf. III,1.10).

Dynamical systems of this type will be termed *differential* ld systems, and the term may also be used in the analogous situation that f in (1) is a continuous map into R^n of a differential n-manifold (cf. I,3). In the present book the term "differential ld system" will have precisely this sense; in particular, if f in (1) is discontinuous but still defines an ld system as in II,1.3.1, then the latter would not be termed differential.

1.4. Lemma. If T *is a continuous lsd system on P, then the escape time* α_x *(cf. II,2.1), considered as a map* $\alpha: P \to R^{\#}$, *is lower semi-continuous;*

$$\alpha_x \leq \liminf_{y \to x} \alpha_y . \tag{2}$$

Conversely, if T *is an abstract lsd system on P and* τ *a topology on P, then (2) is equivalent to 1.1.2.*

Proof. 1.4.1. Take any $x \in P$, $\theta \in [0, \alpha_x)$. Then $(x, \theta) \in$ *domain* T, so that from 1.1.2, $(y, \theta) \in$ *domain* T for y near x; hence $\theta < \alpha_y$ for these y, and

$$\theta \leq \liminf_{y \to x} \alpha_y .$$

Now merely take $\theta \to \alpha_x$ to obtain (2).

1.4.2. For the second assertion, assume the described situation, and take any $(x, \theta) \in$ *domain* T. Then $0 \leq \theta + \varepsilon < \alpha_x$ for small $\varepsilon > 0$; and from (2), for all (x', θ') near (x, θ), one must have $\theta' < \alpha_{x'}$, i.e. $(x', \theta') \in$ *domain* T. This verifies 1.1.2, and completes the proof of 1.4.

1.5. Corollary. If T *is a continuous ld system on P, then the escape times* α_x, β_x *define, respectively, lower and upper semi-continuous maps* $P \to R^{\#}$.

1.6. Corollary. Let T *be a continuous lsd system on P. Then, for any* $\lambda \in R^+$, *the set* $\{x \in P : \lambda < \alpha_x\}$ *is open in P, so that* $\{x \in P : \alpha_x = +\infty\}$ *is a* G_δ *in P. Similarly for continuous ld systems.*

1.7. In differential equation theory, there is a familiar assertion on non-extendability of solutions; for autonomous equations (1) in the situation of 1.3.2, it may be formulated as follows. If $x : (\beta, \alpha) \to G$ is a solution of (1) with $\alpha < +\infty$, and such that it cannot be extended over any larger interval domain (β, α') with $\alpha < \alpha'$, then $x(\theta)$ approaches the frontier of G as $\theta \to \alpha$. This property is almost perfectly reproduced by continuous lsd systems.

1.8. Proposition. Let T *be a continuous lsd system on P, and x a point in P with* $\alpha_x < +\infty$. *Then, for every sequence* $\theta_i \in R^+$ *with* $\alpha_x > \theta_i \to \alpha_x$, $x \mathrm{T} \theta_i$ *does not converge in P. (Similarly for continuous ld systems, and also for lower bounds* β_x.)

(*Proof for ld systems.* $x \top \theta_i$ does not converge in the inherent topology (III,1.2), so that it cannot converge in any weaker topology (III,1.5)).

Proof. This is a variation on the idea used in III,2.8. Assume x, θ_i are as indicated but that $x \top \theta_i \to y$ in P. From 1.4,

$$\alpha_y \leqq \liminf_{z \to y} \alpha_z \leqq \liminf_i \alpha_{x \top \theta_i} = \lim_i (\alpha_x - \theta_i) = 0 ,$$

contradicting $\alpha_y > 0$ in II,2.1.1.

1.9. Corollary. Let $\top$ *be a continuous lsd system on* P, $x \in P$ *with* $\alpha_x < +\infty$. *Then* $x \top \mathsf{R}^+$ *is closed and not contained in any sequentially quasi-compact set. Hence, if* P *is locally sequentially quasi-compact,* $x \top \mathsf{R}^+$ *is unbounded; more precisely,* $\{x \top \theta_n\}$ *is unbounded for every sequence* $\theta_n \in \mathsf{R}^+$ *with* $\alpha_x > \theta_n \to \alpha_x$. *(Similar assertions apply to continuous ld systems.)*

1.10. Corollary. Every continuous dynamical system on a sequentially quasi-compact space is global.

1.11. If $\top$ has the property (stronger than 1.4) that

$$\alpha_x = \lim_{y \to x} \alpha_y \quad \text{for} \quad x \in P ,$$

one obtains a stronger version of 1.8: if $x_i \to x$ in P and $\theta_i \to \alpha_x < +\infty$ in R^+, then $x_i \top \theta_i$ does not converge in P.

Notwithstanding the situation of III,1.6, the above assumption is considerably restrictive, as shown in the following example (however, see V,3.8).

1.12. Example. Take

$$P = \{(x, y) \in \mathsf{R}^2 : x < 0 \text{ or } y > 0\}$$

with its natural topology, and define $\top$ by

$$(x, y) \top \theta = (x + \theta, y)$$

for $(x, y) \in P$, $\theta \in \mathsf{R}^1$, if $(x + \theta, y) \in P$. Then $\top$ is a continuous ld system on P (from 1.3.2 and II,1.3.3), and

$$\alpha_{(x,y)} = \begin{cases} +\infty & \text{for} \quad y > 0 , \\ |x| & \text{for} \quad y \leqq 0 . \end{cases}$$

Hence $\lim \alpha_{(x,y)}$ does not exist, for example for $(x, y) \to (-1, 0)$.

1.13. Example. An abstract ld system may well have a continuous semi-dynamical restriction without itself being continuous. To see this, let

$$P = \{(x, y) \in \mathsf{R}^2 : x > 0 \text{ or } y = 0\} ;$$

for $(x, y) \in P$ and $\theta \in \mathsf{R}^1$ again set

$$(x, y) \,\mathsf{T}\, \theta = (x + \theta, y)$$

if the latter point is in P. It is easily verified that the lsd restriction T' of the ld system T on P is compatible with the subset topology of P inherited from R^2; however, T itself is not compatible with this topology, as may be shown on 1.1.2, or more directly from 1.5 applied to $\beta_{(1,0)}$. This problem will receive further attention in 2.5.

1.14. Lemma. If T *is a continuous ld system on* P, *then each solution* t_x *is continuous, and each translation* T_θ *is continuous with domain open in* P; *similarly for continuous lsd systems.*

Proof. 1.1, 1.6, II,1.4.

2. DYNAMICAL CONCEPTS AND CONTINUITY

This rather lengthy section reviews some consequences of the presence of a compatible topology on the carrier of a dynamical system. The concepts considered are those of Chapter II, with the exception of those appearing in II,3; these latter are treated in the next following section.

2.1. Lemma. Let T *be a continuous lsd system on* P. *To every quasi-compact* $X \subset P$ *there exists an open* $G \supset X$ *and an* $\varepsilon > 0$ *such that* $x \,\mathsf{T}\, \theta$ *is defined for all* $x \in G$ *and* $\theta \in [0, \varepsilon]$. *For continuous ld systems the last relation may be replaced by* $\theta \in [-\varepsilon, \varepsilon]$.

This is a direct consequence of a purely topological lemma.

2.2. Lemma. For $k = 1, 2$, *let* P_k *be topological spaces,* $X_k \subset P_k$ *quasi-compact subsets,* $X_1 \times X_2 \subset G$ *with* G *open in* $P_1 \times P_2$. *Then there exist open* $G_k \subset P_k$ *with*

$$X_1 \times X_2 \subset G_1 \times G_2 \subset G\,.$$

(To apply in 2.1, set $P_1 = P$, $P_2 = \mathsf{R}^+$, $G = \textit{domain}\,\mathsf{T}$, $X_1 \times X_2 = X \times (0)$.)

Proof. First fix any $x \in X_1$. To any $y \in X_2$ there exist open U_{xy}, V_{xy} with

$$(x, y) \in U_{xy} \times V_{xy} \subset G\,.$$

Then $\{V_{xk} \mid y \in X_2\}$ is an open cover of X_2, so that there is a finite subcover to be denoted by $\{V_{xk} \mid k = 1, 2, \ldots, n\}$. Set

$$U_x = \bigcap_k U_{xk}\,, \quad V_x = \bigcup_k V_{xk}\,,$$

obtaining

$$(x) \times X_2 \subset U_x \times V_x = \bigcup_k (U_x \times V_{xk}) \subset \bigcup_k (U_{xk} \times V_{xk}) \subset G\,.$$

Similarly, $\{U_x \mid x \in X_1\}$ constitutes an open cover of X_1, and there is a finite subcover $\{U_k \mid k = 1, 2, \ldots, m\}$. Set

$$G_1 = \bigcup_k U_k, \quad G_2 = \bigcap_k V_k,$$

obtaining open G_1, G_2 with

$$X_1 \times X_2 \subset G_1 \times G_2 = \bigcup_k (U_k \times G_2) \subset \bigcup_k (U_k \times V_k) \subset G$$

as required.

Next there is exhibited an interesting and important property of continuous ld systems, serving as an introduction to 2.4–6.

2.3. Proposition. If T *is a continuous ld system on* P*, then for each open* $G \subset P$ *and arbitrary* $A \subset \mathsf{R}^1$*, the set* $G \,\mathsf{T}\, A$ *is open.*

Proof. Since

$$G \,\mathsf{T}\, A = \bigcup_{\theta \in A} (G \,\mathsf{T}\, \theta)$$

from II,2.13, it is sufficient (and necessary) to show that each $G \,\mathsf{T}\, \theta$ is open. Take any $x \in G \,\mathsf{T}\, \theta$; then $x \,\mathsf{T}\, (-\theta) \in G$ is defined, so that from continuity, $y \,\mathsf{T}\, (-\theta) \in G$ for y near x; hence

$$y = (y \,\mathsf{T}\, (-\theta)) \,\mathsf{T}\, \theta \in G \,\mathsf{T}\, \theta$$

for all y near x, as was to be proved.

Continuous semi-dynamical systems do not, in general, have the property described in 2.3, as shown in 2.6. For definiteness we formulate:

2.4. Definition. A continuous dynamical system T on P is said to have the *openness property* if $G \,\mathsf{T}\, \theta$ is open whenever G is open in P and $\theta \in \mathsf{R}^1$.

Hence 2.3 states that a continuous ld system has the openness property. Observe that for lsd systems the condition in 2.4 need only be required of $\theta \in \mathsf{R}^+$, since for $\theta < 0$, $G \,\mathsf{T}\, \theta$ is void and hence open. The main result may now be stated.

2.5. Theorem. Let T *be an abstract lsd system on a topological space* P*. A necessary and sufficient condition for* T *to be the lsd restriction of a continuous ld system on* P *is that* T *be a continuous lsd system on* P *with the -unicity and openness properties.*

Proof. 2.5.1. Let T' be a continuous ld system on P, and T its lsd restriction. Then T is a continuous lsd system on P: 1.1.1–2 follows by partialization;

$$\textit{domain}\, \mathsf{T} = (P \times \mathsf{R}^+) \cap \textit{domain}\, \mathsf{T}';$$

–unicity from II,2.5; openness from 2.3 applied to T', since $G \,\mathsf{T}\, \theta = G \,\mathsf{T}'\, \theta$ for $\theta \geqq 0$.

2.5.2. Now assume T satisfies the indicated conditions. First, show that there are no *s*-points in P.

Take any $x \in P$, and then open $G \subset P$, $\theta > 0$ such that

$$x \in G\,, \quad G \times [0, \theta] \subset domain\ \mathsf{T}\,.$$

By assumption, $G\,\mathsf{T}\,\theta$ is open and contains $x\,\mathsf{T}\,\theta$; since $x\,\mathsf{T}\,(\theta - \varepsilon) \to x\,\mathsf{T}\,\theta$ as $\varepsilon \to 0^+$,

$$x\,\mathsf{T}\,(\theta - \varepsilon) \in G\,\mathsf{T}\,\theta$$

for a sufficiently small $\varepsilon > 0$. In particular, there is an $x' \in G$ with

$$x\,\mathsf{T}\,(\theta - \varepsilon) = x'\,\mathsf{T}\,\theta\,,$$

and from II,2.6, $x = x'\,\mathsf{T}\,\varepsilon$ with $\varepsilon > 0$. Thus, indeed, no $x \in P$ is an *s*-point.

2.5.3. From the preceding and II,2.9 it now follows that on P there is a (unique) abstract ld system $\mathsf{T}' \supset \mathsf{T}$; and it only remains to prove that T' is a continuous ld system on P. Take $(x, \theta) \in domain\ \mathsf{T}$, and any open $G \subset P$ with $x\,\mathsf{T}'\,\theta \in G$ (we need only consider $\theta \leqq 0$). By assumption,

$$x \in G\,\mathsf{T}\,(-\theta)\,, \quad G\,\mathsf{T}\,(-\theta) \text{ open.}$$

Since x is a non-*s*-point, 2.5.2, there are $x' \in P$, $\varepsilon > 0$ with

$$x'\,\mathsf{T}\,\varepsilon = x\,, \quad x'\,\mathsf{T}\,[0, 2\varepsilon] \subset G\,\mathsf{T}\,(-\theta)\,, \quad (x') \times [0, 2\varepsilon] \subset domain\ \mathsf{T}\,.$$

From 1.2.1′ for T, there exists an open $G' \ni x'$ with

$$G' \times [0, 2\varepsilon] \subset domain\ \mathsf{T}\,, \quad G'\,\mathsf{T}\,[0, 2\varepsilon] \subset G\,\mathsf{T}\,(-\theta)\,.$$

Then $x = x'\,\mathsf{T}\,\varepsilon \in G'\,\mathsf{T}\,\varepsilon$, open by assumption; $(\theta - \varepsilon, \theta + \varepsilon)$ is an open neighbourhood of θ in R^1; and finally

$$(G'\,\mathsf{T}\,\varepsilon)\,\mathsf{T}'\,(\theta - \varepsilon, \theta + \varepsilon) = (G'\,\mathsf{T}\,(0, 2\varepsilon))\,\mathsf{T}'\,\theta \subset (G\,\mathsf{T}\,(-\theta))\,\mathsf{T}'\,\theta = G\,.$$

This proves 1.2.1′ for T' and concludes the proof of 2.5.

Examples. 2.6.1. Take R^+ with its natural topology and gsd system $+$. Then $+$ is continuous with $-$unicity, but without the openness property: $[0, \varepsilon)$ is open in R^+ but $[0, \varepsilon) + 1 = [1, 1 + \varepsilon)$ is not. Of course, R^+ contains an *s*-point, so that this example merely illustrates 2.5.2.

Observe that $+$ has the following *weak openness* property: $G + \mathsf{R}^+$ is open for each open G. Indeed, as basic open sets in R^+ one may take intervals

$$[0, \varepsilon) \text{ and } (\varepsilon', \varepsilon) \text{ with } 0 < \varepsilon' < \varepsilon\,;$$

and then

$$[0, \varepsilon) + \mathsf{R}^+ = \mathsf{R}^+\,, \quad (\varepsilon', \varepsilon) + \mathsf{R}^+ = (\varepsilon', +\infty)$$

are both open in R^+.

2.6.2. As a second example take the continuous gsd system T' of 1.13, and the set

$$G = \{(x, y) \in P : x < 0\} = \{(x, 0) \in \mathsf{R}^2 : x < 0\} .$$

Then G is open in P but neither $G \,\mathsf{T}'\, \mathsf{R}^+$ nor any $G \,\mathsf{T}'\, \theta$ is open in P. Since T' is the lsd restriction of an ld system T on P, 2.5 yields that T is not a continuous ld system on P, in agreement with 1.13.

For closed sets X there is no direct analogue to 2.3: even if X is a singleton and A closed (in an Euclidean carrier), $X \,\mathsf{T}\, A$ need not be closed; and one may even interpret limit set theory as the study of a particular case of this situation. At least there is a weaker result, as follows.

2.7. Lemma. Let T *be a continuous gd system on* P. *If* $X \subset P$ *and* $A \subset \mathsf{R}^1$ *with* A *bounded, then*

$$\overline{X \,\mathsf{T}\, A} = \overline{X} \,\mathsf{T}\, \overline{A} .$$

Thus $X \,\mathsf{T}\, A$ *is closed if* X *is closed and* A *compact.*

Proof. From continuity of the map $\mathsf{T} : P \times \mathsf{R}^1 \to P$ one concludes

$$\overline{X} \,\mathsf{T}\, \overline{A} \subset \overline{X \,\mathsf{T}\, A} .$$

For the converse inclusion, take any $x' \in \overline{X \,\mathsf{T}\, A}$; then there exist $x_i \in X$, $\theta_i \in A$ with $x_i \,\mathsf{T}\, \theta_i \to x'$. Since $\overline{A}$ is compact, one may assume that the θ_i converge in R^1, $\theta_i \to \theta \in \overline{A}$, whereupon

$$x_i = (x_i \,\mathsf{T}\, \theta_i) \,\mathsf{T}\, (-\theta_i) \to x' \,\mathsf{T}\, (-\theta) .$$

Therefore $x' \,\mathsf{T}\, (-\theta) \in \overline{X}$, and

$$x' = (x' \,\mathsf{T}\, (-\theta)) \,\mathsf{T}\, \theta \in \overline{X} \,\mathsf{T}\, \overline{A} ,$$

concluding the proof.

2.8. Lemma. Let T *be a continuous lsd system on* P, *and* $X \subset P$, $A \subset \mathsf{R}^1$.

1. If $0 \in A$ *and both* X *and* A *are (pathwise) connected, then so is* $X \,\mathsf{T}\, A$; *and if* T *is global, the condition* $0 \in A$ *may be omitted.*

2. If both X *and* A *are quasi-compact with* $X \times A \subset$ *domain* T, *then* $X \,\mathsf{T}\, A$ *is quasi-compact; as a partial converse, if* $A = [0, \varepsilon]$ *and both* X, $X \,\mathsf{T}\, A$ *are quasi-compact, then* $X \times A \subset$ *domain* T.

Proof of 1. From the assumptions,

$$X \,\mathsf{T}\, A = X \,\mathsf{T}\, ((0) \cup A) = X \cup (X \,\mathsf{T}\, A) = X \cup \bigcup_{x \in X} (x \,\mathsf{T}\, A)$$

is a decomposition into connected summands, all intersecting the first; hence their union is connected (Bourbaki, 1951, § 11,1). If T is global, then $X \,\mathsf{T}\, A$ is the image of a connected set $X \times A$ under a continuous map T, and thus $X \mathsf{T} A$ is connected (Bourbaki, 1951, § 11,2). Clearly, a similar proof may be carried through for pathwise connectedness.

Proof of 2. If $X \times A \subset \textit{domain}\ \mathsf{T}$ then T partializes to a continuous map onto,

$$\mathsf{T}_p : X \times A \to X \mathsf{T} A ;$$

hence $X \mathsf{T} A$ is quasi-compact if both X, A are (Bourbaki, 1951, § 10).

Finally, let $A = [0, \varepsilon]$, and assume that $X \times A \subset \textit{domain}\ \mathsf{T}$ does not hold. Then $(x) \times A \notin \textit{domain}\ \mathsf{T}$ for some $x \in X$, and from 1.9, $x \mathsf{T} A$ is not contained in any quasi-compact subset of P, in particular $X \mathsf{T} A$ cannot be quasi-compact. This concludes the proof of 2.8.

Concerning the process of relativization applied to continuous dynamical systems, one has the following two results.

2.9. Proposition. If T is a continuous ld system on P and $Q \subset P$ is either open or invariant, then the relativization of T to Q is an ld system compatible with the subset topology of Q.

Proof. In either case Q is *t*-open, from III,1.5 and III,1.2; apply III,1.13.

2.10. Proposition. If T is a continuous lsd system on P and $Q \subset P$ is either open or $+$invariant, then the relativization of T to Q is an lsd system T' compatible with the subset topology; if T has $-$unicity, so does T'. Assuming Q open, the set of s-points of T in Q is precisely the set of s-points of T', and T' has the openness property if T does.

Proof. In the present case, there is no inherent topology available, and the proof needs to be performed from first principles; however this is not at all difficult.

2.10.1. If Q is $+$invariant, then existence of the relativization T' follows from II,5.4. Conditions 1.1.1$-$2 and also the assertion concerning $-$unicity then follow from $\mathsf{T}' \subset \mathsf{T}$ and

$$\textit{domain}\ \mathsf{T}' = (Q \times \mathsf{R}^+) \cap \textit{domain}\ \mathsf{T} .$$

2.10.2. From now on, assume that Q is open. Define T', in analogy with III,1.13, by setting

$$x \mathsf{T}' \theta = x \mathsf{T} \theta \quad \text{iff} \quad x \mathsf{T} \theta' \in Q \quad \text{for all} \quad \theta' \in [0, \theta] .$$

It is then easily shown that T' is an abstract lsd system on Q with $\mathsf{T}' \subset \mathsf{T}$. In particular, T' defines a continuous map, i.e. has 1.1.1, and T' has $-$unicity if T does. To prove 1.1.2 for T', let $(x, \theta) \in \textit{domain}\ \mathsf{T}'$, i.e.

$$(x) \times [0, \theta] \subset \textit{domain}\ \mathsf{T} , \quad x \mathsf{T} [0, \theta] \subset Q .$$

Using 1.2.1′ and 2.2 for T, there exists a set $G \times [0, \theta')$ open in $P \times \mathsf{R}^+$ with

$$(x) \times [0, \theta] \subset G \times [0, \theta') \subset \textit{domain}\, \mathsf{T}\,, \quad G\, \mathsf{T}\, [0, \theta') \subset Q\,.$$

It follows that $G \times [0, \theta')$ is an open subset of *domain* T′, and a neighbourhood of (x, θ); hence *domain* T′ is indeed open in $Q \times \mathsf{R}^+$.

2.10.3. To prove that T′ has the openness property if T does, take any open $G \subset Q$, $\theta \in \mathsf{R}^+$; then

$$G\, \mathsf{T}'\, \theta = G'\, \mathsf{T}'\, \theta \quad \text{with} \quad G' = \{x \in G : x\, \mathsf{T}'\, \theta \text{ defined}\}$$

and G' open in Q, and hence also in P. But then $G'\, \mathsf{T}'\, \theta = G'\, \mathsf{T}\, \theta$ is open in P.

2.10.4. For the assertion concerning *s*-points, let S and S' be the *s*-point sets of T and T′ respectively. From II,3.15 and $\mathsf{T}' \subset \mathsf{T}$, $Q - S' \subset P - S$, so that

$$S \cap Q \subset S'\,.$$

If $x \in Q - S$, for example $x = x'\, \mathsf{T}\, \theta$ with $x' \in P$, $\theta > 0$, then $x'\, \mathsf{T}\, (\theta - \varepsilon) \in Q$ for sufficiently small $\varepsilon > 0$, since Q is open; and then

$$x = (x'\, \mathsf{T}\, (\theta - \varepsilon))\, \mathsf{T}\, \varepsilon = (x'\, \mathsf{T}\, (\theta - \varepsilon))\, \mathsf{T}'\, \varepsilon$$

is a non-*s*-point relative to T′, i.e. $x \in Q - S'$. This completes the proof of 2.10.

2.11. Example. In 2.10 with Q +invariant, if T has the openness property then the relativized system T′ need not have even the weak openness property. Indeed, the gd system on R^2 of II,1.3.3 is a continuous gd system, so that according to 2.5 its semi-dynamical restriction has openness; the gsd system T′ of 1.13 is indeed obtained by relativization of T to its +invariant subset P, and from 2.6.2, T′ does not have the weak openness property.

The following two lemmas consist of results on invariant sets, etc., in the presence of a compatible topology.

2.12. Lemma. Let T *be a continuous lsd system on P, and $X \subset P$ a +invariant subset. Then the closure of X, and also each of its components, is +invariant; and if* T *has the openness property, the interior of X is also +invariant.*

Proof. From continuity, $X\, \mathsf{T}\, \mathsf{R}^+ \subset X$ implies $\overline{X}\, \mathsf{T}\, \mathsf{R}^+ \subset \overline{X\, \mathsf{T}\, \mathsf{R}^+} \subset \overline{X}$, i.e. +invariance of $\overline{X}$. For the assertion concerning connectedness, apply 2.8.1 to $X\, \mathsf{T}\, \mathsf{R}^+ = X$; the last assertion is obvious.

2.13. Lemma. Let T *be a continuous ld system on P. If $X \subset P$ is +invariant, then its closure, interior and each component are all +invariant, and its frontier is t-convex. Analogous results hold for −invariant and invariant sets; furthermore, the frontier of an invariant set is itself invariant.*

Proof. The first assertions follow from 2.12 and 2.3; if X is $+$invariant, then its frontier

$$\operatorname{Fr} X = \bar{X} - \operatorname{Int} X \tag{1}$$

is the difference of $+$invariant sets, and hence t-convex. For results on $-$invariant sets apply 2.12 and orientation change, also obtaining the assertions for invariant sets. Finally, if X is invariant, its frontier is the difference of invariant sets (1) and thus is invariant.

2.14.1. Example. In 2.12, the closure of an invariant set need not be $-$invariant. To see this, consider the set

$$X = \{(x, y) \in P : y < 0\} = \{(x, y) \in \mathsf{R}^2 : y < 0 < x\}$$

in the situation of example 1.13. Then X is invariant relative to T', but

$$\bar{X} = \{(x, y) \in P : y \leqq 0 \text{ or } x = y = 0\}$$

is not $-$invariant, as

$$(-1, 0) \in P - \bar{X}, \quad (-1, 0)\, \mathsf{T}'\, 1 = (0, 0) \in \bar{X}.$$

2.14.2. Example. In 2.13, a set X with t-convex frontier is often neither $+$invariant nor $-$invariant. As an example, in R^1 with its natural topology and gd system, each singleton is t-convex and coincides with its frontier.

A slightly more involved example is the following. Consider the continuous gd system defined on R^2 by the differential equation

$$\frac{dz}{d\theta} = |\operatorname{Re} z|;$$

its critical point set is the imaginary axis. For X take the interior of the unit circle in R^2 (cf. Fig. 1). Then X is simply pathwise connected and regularly open, its frontier is t-convex, but X is neither $+$invariant nor $-$invariant.

The continuous analogue to II,5.16 will also be useful later.

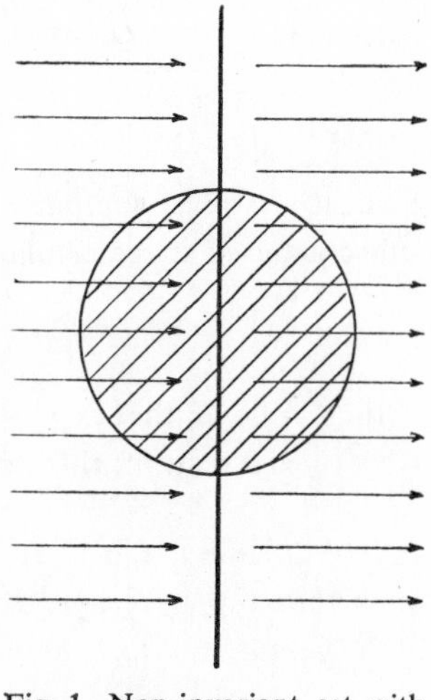

Fig. 1. Non-invariant set with t-convex frontier.

2.15. Lemma. Let T_0 be a continuous ld system on P_0, let $P \subset P_0$ be open and t-convex in P_0, and T the relativization of T_0 to P. Then, to any continuous morphism

$$f : (P, \mathsf{T}) \to (P', \mathsf{T}') \text{ in } \mathscr{D}_l$$

with T' a continuous gd system on P', there exists a unique continuous morphism g in $\mathscr{D}_l$ extending f and defined on the subset generated by P in P_0.

Proof. Existence and unicity of the abstract extension g was proved in II,5.16; it remains to verify that this g is continuous. Recall that g is defined by

$$g(x \mathsf{T}_0 \theta) = f(x) \mathsf{T}' \theta \quad \text{for} \quad x \in P, \theta \in \mathsf{R}^1$$

whenever $x \mathsf{T}_0 \theta$ is defined, and take any open neighbourhood G' of $g(x \mathsf{T}_0 \theta)$ in P'. Then

$$G = \{y \in P : f(y) \mathsf{T}' \theta \in G'\}$$

is an open neighbourhood of x in P so that, from 2.3, $G \mathsf{T}_0 \theta$ is an open neighbourhood of $x \mathsf{T}_0 \theta$ in P_0. Obviously now

$$g(G \mathsf{T}_0 \theta) = f(G) \mathsf{T}' \theta \subset G' ;$$

this concludes the proof.

In conclusion, the sections in troducedin II,6 will be treated; for further results also see V,2.

2.16. *Lemma. Let* T *be a continuous lsd system on* P, *and* Q *a quasi-compact subset of* P; *if* Q *is a section locally at each point* $x \in Q$, *then* Q *is a section.*

Proof. From the assumptions it follows that there exists a finite open cover $\{G_k \mid 1 \leqq k \leqq n\}$ of Q and an $\varepsilon > 0$ such that each $Q \cap G_k$ is a section of t-extent ε. Now assume that Q is not a section; then there are $x_i \in Q$, $\theta_i \in \mathsf{R}^+$ with

$$0 < \theta_i \to 0 , \quad x_i \in Q \ni x_i \mathsf{T} \theta_i .$$

Since Q is quasi-compact, one may assume that $x_j \to x \in Q$ and $\theta_j \to 0$ for some subsequences; in particular, $x \in G_k$ for some member G_k of the open cover. From

$$x_j \to x , \quad x_j \mathsf{T} \theta_j \to x \mathsf{T} 0 = x$$

it then follows that $x_j \mathsf{T} \theta_j \in Q \cap G_k \ni x_j$ for large j, contradictiong the fact that $Q \cap G_k$ is a section; this proves the assertion.

2.17. *Lemma. Let* T *be a continuous ld system on* P, Q *a closed subset,* $\varepsilon > 0$ *such that* $I = [-\frac{1}{2}\varepsilon, \frac{1}{2}\varepsilon]$ *has* $Q \times I \subset$ domain T. *Then* Q *is a section of t-extent* ε *iff* T *partializes to a homeomorphism* $\mathsf{T}_p = \mathsf{T} \mid (Q \times I)$,

$$\mathsf{T}_p : Q \times I \approx Q \mathsf{T} I .$$

Proof. In any case, from the assumptions and $\mathsf{T}_p \subset \mathsf{T}$ it follows that T_p is a continuous map onto $Q \mathsf{T} I$; from II,6.8, Q is a section of t-extent ε iff T_p is $1-1$. Thus it only remains to show that the mapping inverse to T_p is continuous if Q is a section as indicated. Let

$$x_i, x \in Q , \quad \theta_i, \theta \in I , \quad x_i \mathsf{T} \theta_i \to x \mathsf{T} \theta .$$

Since I is compact, one may assume $\theta_i \to \theta' \in I$, whereupon

$$x_i = (x_i \top \theta_i) \top (-\theta_i) \to (x \top \theta) \top (-\theta') = x \top (\theta - \theta') .$$

As $|\theta - \theta'| \leqq |\theta| + |\theta'| \leqq \varepsilon$, from the assumption on Q one has that $x \top (\theta - \theta') \in \in Q$, and then $\theta = \theta'$, implying

$$\theta_i \to \theta , \quad x_i \to x .$$

This concludes the proof.

2.18. In the situation of 2.17, Q is essentially a strong deformation retract of $Q \times I$ (Hu, 1959); it is then readily surmised that the same holds of its homeomorph $Q \top I \approx Q \times I$. A retracting deformation $r : Q \top I \to Q$ can be obtained by composing the mappings in

$$Q \top I \xrightarrow{h} Q \times I \xrightarrow{p} Q ,$$

with $h = \top_p^{-1}$ and p the projection map; this retraction is then described more directly by

$$r(x \top \theta) = x \quad \text{for} \quad x \in Q, \ |\theta| \leqq \tfrac{1}{2}\varepsilon .$$

3. TOPOLOGICO-DYNAMICAL CATEGORIES

3.1. One may define, in a natural manner, categories of sets endowed with topological and dynamical structures inter related as in 1.1. As an example, the formulation of 2.15 might be unburdened by introducing the category $\mathscr{D}_l^T$ of continuous ld systems on topological spaces. Similarly one may introduce the category $\mathscr{D}_{sl}^T(P)$ of continuous lsd systems on a fixed abstract set P with various topologies, the category $\mathscr{D}_g(P, \tau)$ of continuous gd systems on a given topological space (P, τ), the category $\mathscr{D}_l^{T_2}$ of continuous ld systems on T_2 spaces, etc. (these are intended to suggest a systematic notation).

For definiteness, the category $\mathscr{D}_l^T$ is introduced as follows. Its objects are triples $(P, \top, \tau)$ with $\top$ an ld system on P and τ a topology on P compatible with $\top$ in the sense of 1.1. (Again, the first term, P, is superfluous.) The morphisms of $\mathscr{D}_l^T$,

$$f : (P, \top, \tau) \to (P', \top', \tau') ,$$

correspond to set-theoretical maps $f : P \to P'$, continuous in the indicated topologies, and with

$$f(x \top \theta) = f(x) \top' \theta \tag{1}$$

whenever $x \top \theta$ is defined.

Omission of the third term in any object $(P, \top, \tau) \in \mathscr{D}_l^T$ yields the object $(P, \top) \in \mathscr{D}_l$, thereby defining an *amnestic* functor $\mathscr{D}_l^T \to \mathscr{D}_l$.

3.2. Lemma. Adjunction of the inherent topology to an ld system $(P, \mathrm{T}) \in \mathscr{D}_l$ *defines a functor* $I : \mathscr{D}_l \to \mathscr{D}_l^T$, *which is a left inverse to the amnestic functor in the sense that commutativity obtains in*

$$\begin{array}{ccc} \mathscr{D}_l & = & \mathscr{D}_l \\ & {}_{I}\searrow \quad \nearrow_{\textit{amnestic}} & \\ & \mathscr{D}_l^T & \end{array} \tag{2}$$

Proof. For any object $(P, \mathrm{T}) \in \mathscr{D}_l$, let $\tau(\mathrm{T})$ be the inherent topology associated with T according to III,1.1; $(P, \mathrm{T}, \tau(\mathrm{T})) \in \mathscr{D}_l^T$ then follows from III,1.3. For any morphism $f : (P, \mathrm{T}) \to (P', \mathrm{T}')$ in $\mathscr{D}_l$ let $f : (P, \mathrm{T}, \tau(\mathrm{T})) \to (P', \mathrm{T}', \tau(\mathrm{T}'))$ correspond to the same set-theoretical map $f : P \to P'$ (the terminology is as in II,3.9). To prove $f \in \mathscr{D}_l^T$, it is necessary to verify that f is continuous relative to the topologies $\tau(\mathrm{T})$ and $\tau(\mathrm{T}')$. From III,1.4 and Bourbaki (1951, I, § 7), this obtains iff ft_x is continuous for each solution t_x of T. Now, (1) may be rewritten as

$$ft_x \subset t'_{f(x)}$$

with t'_y denoting solutions of T', and hence ft_x is continuous since $t'_{f(x)}$ is such. It follows that f is continuous, and therefore there indeed results a functor $\mathscr{D}_l \to \mathscr{D}_l^T$ as indicated.

3.3. In II,3 there were exhibited the "categorial constructions" of direct products, direct sums and factor systems in the abstract dynamical categories. It is then natural to ask, for example, whether the abstract direct product of continuous dynamical systems is compatible with the product topology of the carrier. The answers are almost as reasonable as may be expacted.

3.4. Proposition. Let $(P_i, \mathrm{T}_i, \tau_i) \in \mathscr{D}_l^T$ *for* $i \in I$; *if the direct product* (P, T) *of abstract* (P_i, T_i) *exists in* $\mathscr{D}_l$, *then* T *is compatible with the product topology* τ *of the* τ_i, *i.e.* $(P, \mathrm{T}, \tau) \in \mathscr{D}_l^T$.

A similar result holds in $\mathscr{D}_{sl}^T$; *and relations between openness properties of* T_i, T *are then as follows.*

1. If, for each $i \in I$, *escape times in* (P_i, T_i) *are unbounded, then all* T_i *have the openness property if* T *does and* $P \neq \emptyset$.

2. If the index set I is finite and all T_i *have the openness property, then so does* T.

Proof. 3.4.1. Consider the semi-dynamical case, and assume that the direct product (P, T) of the (P_i, T_i) exists in $\mathscr{D}_{sl}$; thus, from II,3.2. there is a finite $J \subset I$ such that all T_i with $i \notin J$ are global.

3.4.2. That T is compatible with τ will be proved by verifying 1.2.1′. Take any

$x = [x_i] \in P$, $\theta \in \mathsf{R}^+$, open $G \subset P$ such that $x \mathsf{T} \theta \in G$. However, for G it suffices to take a sub-basic open set, i.e. of the form

$$G = G_j \times P^{(j)}$$

for some $j \in I$, G_j open in P_j, and with $P^{(j)}$ denoting the set

$$P^{(j)} = \prod_{i \neq j} P_i ;$$

also, assume $j \notin J$ (in opposite case the necessary modifications are rather obvious). Then $x_j \mathsf{T}_j \theta \in G_j$, so that 1.2.1′ applied to T_j yields open $H_j \subset P_j$, open $A_j \subset \mathsf{R}^+$ with

$$(x_j, \theta) \in H_j \times A_j \subset \mathit{domain}\, \mathsf{T}_j \,, \quad H_j \mathsf{T}_j A_j \subset G_j \,.$$

Also perform this for each $i \in J$, and set

$$H = H_j \times \prod_{i \in J} H_i \times \prod_{j \neq i \notin J} P_i \,, \quad A = A_j \cap \bigcap_{i \in J} A_i \,.$$

Then obviously

$$(x, \theta) \in H \times A \subset \mathit{domain}\, \mathsf{T} \,, \quad H \mathsf{T} A \subset G \,,$$

as it was required to prove. For ld systems (i.e. in the category $\mathscr{D}_l^T$) a similar proof may be used, merely replacing R^+ by R^1.

3.4.3. Next, assume that the product system has the openness property. Take any $j \in I$, G_j open in P_j, $\theta \in \mathsf{R}^+$, and attempt to prove that $G_j \mathsf{T}_j \theta$ is open, i.e. that T_j has openness. With notation as in 3.4.2, set

$$G = G_j \times P^{(j)} ;$$

then G is a sub-basic open set in P, and $G \mathsf{T} \theta$ is open by assumption; since the projection map $p_j : P \to P_j$ is interior (Bourbaki, 1951, I, § 8), it suffices to prove that

$$G_j \mathsf{T}_j \theta = p_j(G \mathsf{T} \theta) \,.$$

Since $G_j = p_j(G)$ by construction, the morphism condition for p_j (cf. (2) in II,3.1) yields

$$p_j(G \mathsf{T} \theta) \subset G_j \mathsf{T}_j \theta \,.$$

For the converse inclusion, take any $x_j \in G_j$ with $x_j \mathsf{T}_j \theta$ defined; applying the assumption in 1, for all $i \neq j$ take $x_i \in P$ with $x_i \mathsf{T}_i \theta$ defined, and set $x = [x_i]$. Then $x \in G$, $x \mathsf{T} \theta$ is defined, so that

$$x \mathsf{T} \theta \in G \mathsf{T} \theta \,, \quad p_j(x \mathsf{T} \theta) = x_j \mathsf{T}_j \theta \,;$$

this proves $G_j \mathsf{T}_j \theta \subset p_j(G \mathsf{T} \theta)$, and concludes the proof of 3.4.1.

3.4.4. Finally, assume that I is finite and all T_i have the openness property; to prove openness for T it suffices to consider only basic open sets of the form

$$G = \Pi G_i\,, \quad G_i \text{ open in } P_i\,.$$

Take any $\theta \in \mathsf{R}^1$; then, from the definition of the product system,

$$G \,\mathsf{T}\, \theta = \Pi(G_i \,\mathsf{T}_i\, \theta)\,,$$

and this is indeed an open set in P. This completes the proof of 3.4.

Examples. 3.5.1. Several of the examples in the present chapter are instances of the following construction. On R^1 with its natural topology take two continuous gd systems, the natural and the trivial; then form the direct product of these two objects, in accordance with II,3.2 and 3.4:

$$(x, y) \,\mathsf{T}\, \theta = (x \,\mathsf{T}_1\, \theta, y \,\mathsf{T}_2\, \theta) = (x + \theta, y)\,.$$

The resulting object is a continuous gd system on R^2 (with its natural topology; of course, this may be verified directly, or by applying 1.3.2 to II,1.3.3). The mentioned examples are then obtained by restriction or relativization.

3.5.2. In 3.4, if all T_i have the openness property but the set I is infinite, then the direct product need not have openness; this is easily obtained from the example of II,3.3.2 by using 2.5.

3.6. Lemma. Let $(P_i, \mathsf{T}_i, \tau_i) \in \mathscr{D}_l^T$ for $i \in I$ with the P_i disjoint. Then the abstract direct sum (P, T) of the (P_i, T_i) has T compatible with the direct sum topology τ, i.e. $(P, \mathsf{T}, \tau) \in \mathscr{D}_l^T$. A similar results holds in $\mathscr{D}_{sl}^T$; furthermore, T has the openness property iff all T_i do.

Proof. II,3.4; recall that a set $G \subset P$ is open relative to τ iff, for each $i \in I$, $G \cap P_i$ is open relative to τ_i (Bourbaki, 1951, I, § 8).

3.7. Example. The natural continuous gsd system $+$ on R^1 relativizes to continuous lsd systems T_1 and T_2 on

$$(-\infty, 0) \quad \text{and} \quad [0, +\infty) \tag{3}$$

respectively (cf. 2.10). The lsd system T' of II,2.11 is their direct sum, as noted in II,3.4; according to 3.6, T' is a continuous lsd system on the topological direct sum of the carriers (3) (this latter is homeomorphic to $(-\infty, -1) \cup (0, +\infty)$). Note that T' is not compatible with the natural topology of R^1.

3.8. Proposition. Let $(P, \mathsf{T}, \tau) \in \mathscr{D}_l^T$, and let $\sim$ be an equivalence relation on P, strongly compatible with T. On $P' = P/\!\sim$ as carrier, let T' be the factor ld system of (P, T) modulo $\sim$, and τ' the quotient topology of (P, τ) modulo $\sim$; then T' is *compatible with* τ', *i.e.* $(P', \mathsf{T}', \tau') \in \mathscr{D}_l^T$. *A similar results holds in* $\mathscr{D}_{sl}^T$, *and if* $\sim$ *is an open equivalence relation, then* T' *has the openness property if* T *does.*

(For the terms used here, refer to II,3.5–6, 2.4 and Bourbaki (1951, I, § 9).)

Proof. 3.8.1. This proceeds by verifying 1.2.1′ for T' and τ', and is essentially the same for ld and lsd systems; for definiteness, consider the latter.

Take any $x' \in P'$, $\theta' \in \mathsf{R}^+$, open $G' \subset P'$ with $x' \mathsf{T}' \theta' \in G'$. Set $G = e^{-1}(G')$, an open $\sim$saturated set in P (as in II,3.6, $e : P \to P'$ is the canonic quotient map, $x \in e(x)$), and select an $x \in e^{-1}(x')$ in P with $x \mathsf{T} \theta'$ defined. Then $x \mathsf{T} \theta' \in G$, so that from continuity of T, one may take a compact neighbourhood A of θ' in R^+ with

$$(x) \times A \subset \mathit{domain}\ \mathsf{T}, \quad x \mathsf{T} A \subset G. \tag{4}$$

Next, set

$$H = \{y \in P : (y) \times A \subset \mathit{domain}\ \mathsf{T},\ y \mathsf{T} A \subset G\}, \tag{5}$$

so that $x \in H$ from (4).

3.8.2. Now prove that H is $\sim$saturated and open in P. If

$$z \sim y \in H, \quad \theta \in A,$$

then $y \mathsf{T} \theta$ is defined, hence $z \mathsf{T} \theta$ is defined and $z \mathsf{T} \theta \sim y \mathsf{T} \theta$ from the strong compability assumption. Since $G = e^{-1}(G')$ is $\sim$saturated,

$$z \mathsf{T} \theta \sim y \mathsf{T} \theta \in G \quad \text{implies} \quad z \mathsf{T} \theta \in G;$$

and, as $\theta \in A$ was taken arbitrarily, this proves $z \in H$, i.e. H is $\sim$saturated.

Second, assume

$$H \not\ni y_i \to y \in H. \tag{6}$$

Using 2.2,

$$(y_i) \times A \subset \mathit{domain}\ \mathsf{T}$$

for large i, so that $y_i \notin H$ requires that

$$y_i \mathsf{T} \theta_i \notin G \quad \text{for some} \quad \theta_i \in A.$$

Since A was taken compact, one may assume $\theta_j \to \theta \in A$ for some subsequence, whereupon

$$y_j \mathsf{T} \theta_j \to y \mathsf{T} \theta \in G$$

from the assumption $y \in H$. However, this contradicts openness of G; hence (6) is impossible, and H is open. Summarizing, H is an $\sim$invariant open neighbourhood of x in P.

3.8.3. By definition of the quotient topology, $e(H)$ is then open in P'. Now, take an open neighbourhood A' of θ' in R^+ with $A' \subset A$. Then, using (5),

$$e(H) \mathsf{T}' A' = e(H \mathsf{T} A') \subset e(H \mathsf{T} A) \subset e(G) = G' .$$

This completes the proof that T' and τ' satisfy 1.2.1′.

3.8.4. Finally, assume that $\sim$ is an open equivalence relation and that T has the openness property; and take any G' open in P', $\theta' \in \mathsf{R}^+$. Then

$$e : (P, \tau) \to (P', \tau')$$

is an interior mapping, and $e^{-1}(G') \mathsf{T} \theta'$ is open, so that

$$G' \mathsf{T} \theta' = e(e^{-1}(G') \mathsf{T} \theta')$$

is indeed open. This completes the proof of 3.8.

3.9. Remark. In comparing 3.8 with II,3.6, in the continuous case strong compatibility is required instead of compatibility; concerning this, see examples 3.10 below. If, however, T is global, then compatibility is equivalent with strong compatibility; it is this special case of 3.8 which forms its most important application in V,2.

Examples. 3.10.1. In 3.8, if $\sim$ is merely compatible with T, then T' may fail to be compatible with τ'. To see this, consider the gd system on R^2 as in 3.5.1, and relativize it to the left open semi-plane P,

$$P = \{(x, y) \in \mathsf{R}^2 : x < 0\} ;$$

in accordance with 2.9 there results a continuous ld system T on P.

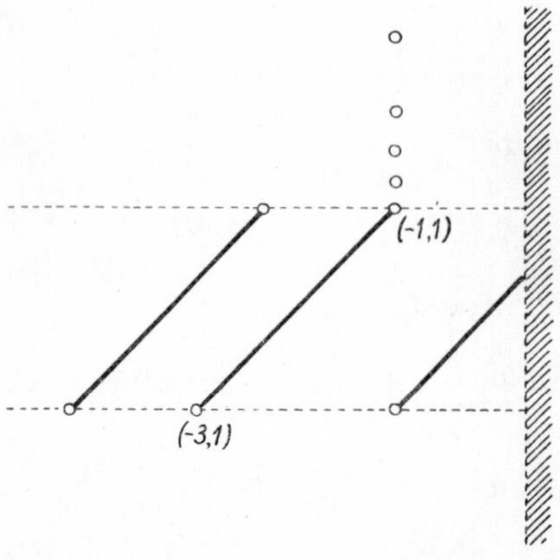

Fig. 2. Carrier of continuous lsd system with discontinuous factor-system: full lines are equivalence classes.

Now define an equivalence relation $\sim$ on P thus: $(x, y) \sim (x', y')$ if either $(x, y) = (x', y')$ or

$$x - y = x' - y' ,$$
$$|y| \leqq 1 \geqq |y'| , \quad x < 0 > x' ,$$

(some equivalence classes are sketched in Fig. 2, which also indicates that $\sim$ is indeed compatible with T). However, the factor system T' is not compatible with the quotient topology, since escape times relative to T' are not lower semi-continuous as they should be according to 1.4:

$$(-1, 1) \sim (-3, -1) ,$$
$$\alpha'_{(-1,1)} = \alpha'_{(-3,-1)} = 3 ,$$
$$\alpha'_{(-1,y)} = 1 \quad \text{for} \quad y \to 1+ .$$

As an aid to visualization, any equivalence relation on P strongly compatible with T must have its equivalence classes "vertical" in fig. 1.

3.10.2. In contrast, strong compatibility of $\sim$ with T is not a necessary condition for the main conclusion of 3.8, viz. compatibility of T' with τ'. Indeed, consider the natural gd system on R^1, relativized to a continuous ld system T on the negative semi-axis, i.e. on

$$P = (-\infty, 0);$$

and also the equivalence relation described by $x \sim y$ in P iff $x - y$ is an integer. Obviously, the quotient space is $P/\sim\, = \mathsf{S}^1$, and the factor system T' is again "natural":

$$x\,\mathsf{T}'\,\theta = xe^{i\theta} \quad \text{for} \quad x \in \mathsf{S}^1,\ \theta \in \mathsf{R}^1$$

(with S^1 identified with the unit circle in the complex plane). In particular, T' is compatible with the quotient topology; however, $\sim$ is not strongly compatible with T.

3.10.3. Let there be given a continuous lsd system T on a topological space P. Among the equivalence relations on P strongly compatible with T there is a maximal one, described, of course, by

$$x \sim y \quad \text{iff} \quad \alpha_x = \alpha_y\,. \tag{7}$$

According to 3.8 one may then construct the corresponding continuous factor system on the quotient space. In a special case there is a satisfactory model for this latter object; this will now be described.

First, consider the continuous ld system defined on R^1 by the autonomous equation $dx/d\theta = x^2$; take its semi-dynamical restriction relativized to R^+, obtaining a continuous lsd system T_0,

$$x\,\mathsf{T}_0\,\theta = \frac{x}{1-\theta x} \quad \text{for} \quad x \in \mathsf{R}^+,\quad 0 \leqq \theta < \frac{1}{x}\,.$$

Next take any continuous lsd system T on P with the property that escape times α_x define a continuous map $\alpha : P \to \mathsf{R}^\#$; and set

$$f(x) = \begin{cases} 0 & \text{if} \quad \alpha_x = +\infty\,, \\ 1/\alpha_x & \text{if} \quad \alpha_x < +\infty\,, \end{cases}$$

obtaining a continuous map $f : P \to \mathsf{R}^+$. From II,2.2 it follows easily that $f : (P, \mathsf{T}) \to$ $\to (\mathsf{R}^+, \mathsf{T}_0)$ in $\mathscr{D}^T_{sl}$, and obviously f induces a morphism $f' : (P', \mathsf{T}') \to (\mathsf{R}^+, \mathsf{T}_0)$ in $\mathscr{D}^T_{sl}$, where (P', T') is the factor system of (P, T) modulo the equivalence relation of (7); this f' maps P' homeomorphically onto R^+ (unless T is global, in which case, of course, P' is a singleton).

3.11. The restriction functor $R : \mathscr{D}_l \to \mathscr{D}_{slu}$ of II,3.12 partializes to a functor $\mathscr{D}_l^T \to \mathscr{D}_{slu}^T$, as noted in 2.5.1. To show that, similarly, the orientation changing functor $C : \mathscr{D}_l \to \mathscr{D}_l$ partializes to a functor $\mathscr{D}_l^T \to \mathscr{D}_l^T$, use the homeomorphism $P \times \mathsf{R}^1 \approx P \times \mathsf{R}^1$ defined by the assignment $(x, \theta) \to (x, -\theta)$.

The process of forming the direct product of two ld systems may be described as the action of a covariant functor with two entries, $\mathscr{D}_l \times \mathscr{D}_l \to \mathscr{D}_l$; 3.4 then states that this functor partializes to a second covariant functor $\mathscr{D}_l^T \times \mathscr{D}_l^T \to \mathscr{D}_l^T$. Similar remarks apply to the direct sum construction.

3.12. The continuous analogue to the remark in II,3.16 may be formulated, inaccurately but perhaps vividly, by saying that the concept of a continuous dynamical system is topologically invariant. Explicity, let T be a continuous lsd or ld system on a topological space P, and $f : P \approx P'$ a homeomorphism; then the dynamical system T' induced by T and f on P', and defined by

$$x' \,\mathsf{T}'\, \theta = f(f^{-1}(x') \,\mathsf{T}\, \theta)\,,$$

is a continuous dynamical system on P'.

The preceding construction may be generalized in a significant manner; this is specific to continuous dynamical systems, there being no parallel applying to abstract systems. Assume given two topological spaces P and P' with P connected and locally pathwise connected, and also a continuous map onto, $f : P \to P'$; then P is called a covering space of P' relative to f if each $x' \in P'$ has a "canonic" connected open neighbourhood G' on P' such that each component of $f^{-1}(G')$ is open in P and mapped homeomorphically onto G' by f (Hu, 1959, III,16).

3.13. Proposition. Let T' be a continuous lsd system on P', and P a covering space of P' relative to $f : P \to P'$. Then there exists a continuous lsd system T on P such that

$$f : (P, \mathsf{T}) \to (P', \mathsf{T}') \text{ in } \mathscr{D}_{sl}^T\,.$$

Furthermore,

$$\alpha_x = \alpha'_{f(x)} \quad \textit{for all} \quad x \in P\,;$$

T has $-$unicity iff T' does; $x \in P$ is an s-point relative to T iff $f(x)$ is an s-point relative to T' (and in the positive case $f^{-1} f(x)$ consists of s-points); T has the openness property iff T' does. A similar result holds in $\mathscr{D}_l^T$.

Proof. 3.13.1. To any path $p' : [0, \theta] \to P'$ and any $x \in P$ with $f(x) = p(0)$ there exists in P a unique "covering" path $p : [0, \theta] \to P$ such that

$$p' = fp\,,$$

(Hu, 1959, III,15). Now apply this result to $+$solutions $t'_{x'} : [0, \alpha'_{x'}) \to P'$ of T', obtaining the following result. With every $x \in P$ there is associated a continuous map

$$t_x : [0, \alpha'_{f(x)}) \to P$$

uniquely determined by $t'_{f(x)} = ft_x$; hence, easily, the partial operator $\top$ defined by

$$x \top \theta = t_x(\theta)$$

for $x \in P$, $\theta \in [0, \alpha'_{f(x)})$, is an lsd system on P, evidently with bounds as asserted. There follows immediately that

$$f(x \top \theta) = f\, t_x(\theta) = t'_{f(x)}(\theta) = f(x) \top' \theta\,,$$

so that f is a morphism in $\mathscr{D}_{sl}$.

3.13.2. Next, verify that $\top$ satisfies condition 1.2.1′. Take $x \in P$, $\theta \in \mathsf{R}^+$, open G in P with $x \top \theta \in G$. Then $f(x) \top' \theta$ is defined and within $f(G)$, with $f(G)$ open in P' (at least for small G). Hence there exists an open $G' \times A \subset P' \times \mathsf{R}^+$ with

$$(f(x), \theta) \in G' \times A \subset \textit{domain}\ \top'\,, \quad G' \top A \subset f(G)\,,$$

whereupon

$$(x, \theta) \in H \times A \subset \textit{domain}\ \top\,, \quad H \top A \subset G$$

for the open component H of $f^{-1}(G')$ containing x (at least for small G), as asserted. In particular, f is a morphism in $\mathscr{D}_{sl}^T$.

3.13.3. Finally, consider the assertions concerning $-$unicity, s-points and openness. Take $x \in P$, $x' \in P'$ with $f(x) = x'$ (either x or x' may be taken arbitrarily); and canonic open neighbourhoods G, G' of x, x' respectively, with $f \mid G$ a homeomorphism $G \approx G'$. From 2.9, the two lsd systems relativize, over the indicated open sets, to systems isomorphic in $\mathscr{D}_{sl}^T$ via the map $f \mid G$; hence if one has $-$unicity, etc., then so does the other. Furthermore, one may apply these results and theorem 2.5 to obtain the assertion concerning ld systems. This completes the proof of 3.13.

4. LIMIT SETS, ORBITAL STABILITY

There will now be introduced a basic and classical concept of dynamical system theory, the limit sets.

4.1. Definition. Let $\top$ be a continuous lsd system on P, and fix an $x \in P$. A point $y \in P$ is called a *+limit point* (of x or of $x \top \mathsf{R}^+$) if there exist $\theta_i \in \mathsf{R}^+$ with

$$\theta_i \to \alpha_x\,, \quad x \top \theta_i \to y\,.$$

The set of +limit points is termed the *+limit set* (of x or $x \top \mathsf{R}^+$) and denoted by L_x^+. For continuous ld systems one also defines *−limit points* and *sets* L_x^- by orientation change, and *limit sets* L_x by $L_x = L_x^+ \cup L_x^-$.

As an example, from 4.1 there follows immediately

4.2. Lemma. If T *is a continuous lsd system on* P *and* $x \in P$ *is either critical or feebly critical or on a cycle, then* $L_x^+ = x \mathsf{T} \mathsf{R}^+$.

Note that, in the inherent topology of an ld system, there are no further limit points than those described in 4.2.

4.3. Proposition. Let T *be a continuous lsd system on* P, *and* $x \in P$. *If* $\alpha_x < +\infty$ *then* $L_x^+ = \emptyset$. *If* $\theta_n \to +\infty$ *are taken arbitrarily in* R^+, *then the Lefschetz formula holds*

$$L_x^+ = \bigcap_{n=1}^{\infty} \overline{x \mathsf{T} [\theta_n, +\infty)}\,. \tag{1}$$

Proof. The first assertion is merely a reformulation of 1.8. Hence, in proving (1), one may assume $\alpha_x = +\infty$. Take $\theta_n \to +\infty$ in R^+. If $y \in L_x^+$, there exist $\theta_i' \to +\infty$ with $x \mathsf{T} \theta_i' \to y$; then $\theta_n < \theta_i'$ for any n and large i, and thus

$$x \mathsf{T} \theta_i' \in x \mathsf{T} [\theta_n, +\infty)\,, \quad y \in \overline{x \mathsf{T} [\theta_n, +\infty)}$$

for all n, proving one inclusion in (1). Conversely, if y is in the set-meet of (1), then any neighbourhood V of y contains some

$$x \mathsf{T} \theta(n, V) \in V\,, \quad \theta_n \leqq \theta(n, V)\,.$$

The $\theta(n, V)$ constitute a generalized sequence with indices (n, V) ordered naturally,

$$(n, V) \leqq (n', V') \quad \text{iff} \quad n \leqq n' \quad \text{and} \quad V \supset V'\,.$$

This implies $\theta_n \leqq \theta(n, V) \to +\infty$ and $x \mathsf{T} \theta(n, V) \to y$ as required in 4.1. Hence, the second inclusion to (1), concluding the proof.

4.4. Corollary. Each $+$*limit set is* $+$*invariant, closed, and with a countable dense subset; a singleton* $+$*limit set is critical.*

Proof. 4.3, 2.12, II,5.6.2–4; since $x \mathsf{T} \mathsf{R}^+$ has a countable dense subset, so does $\overline{x \mathsf{T} \mathsf{R}^+}$ and hence also $L_x^+ \subset \overline{x \mathsf{T} \mathsf{R}^+}$.

4.5. Corollary. $L_x^+ = L_y^+$ *if* x, y *are t-equivalent; if* $y \in L_x^+$ *then* $L_y^+ \subset L_x^+$.

4.6. Lemma. Let T *be a continuous ld system; then each* $+$*limit set (or* $-$*limit, limit) is closed invariant.*

Proof. Having 4.4 it remains to prove $-$invariance for a $+$limit set L_x^+. If $y \mathsf{T} \theta \in L_x^+$ with $\theta \in \mathsf{R}^+$, then $x \mathsf{T} \theta_i \to y \mathsf{T} \theta$ for some $\theta_i \to +\infty$ (cf. 4.3), whereupon

$$x \mathsf{T} (\theta_i - \theta) \to (y \mathsf{T} \theta) \mathsf{T} (-\theta) = y$$

and $y \in L_x^+$ as required. (For continuous gd systems another proof follows from 4.3 and 2.7.)

4.7. Theorem. Let T *be a continuous lsd system on* P*, and assume that a +trajectory* $x \mathsf{T} \mathsf{R}^+$ *is in some quasi-compact part of* P*. Then* L_x^+ *is non-void quasi-compact, and* $\alpha_y = +\infty$ *for all* $y \in L_x^+$*; if, furthermore,* P *is a* T_2 *space, then* L_x^+ *is connected.*

Proof. From 4.3,

$$L_x^+ = \bigcap_n \overline{x \mathsf{T} [n, +\infty)}\,.$$

The sets $\overline{x \mathsf{T} [n, +\infty)}$ constitute a decreasing sequence of non-void quasi-compact sets, and hence their intersection L_x^+ is also non-void quasi-compact (Bourbaki, 1951). According to 4.5, $y \in L_x^+$ then implies that $y \mathsf{T} \mathsf{R}^+$ is contained in the quasi-compact set L_x^+; hence $\alpha_y = +\infty$ from 1.9. If P is a T_2 space, then L_x^+ is a compact non-void limit of compact connected sets, and hence is itself connected (Kuratowski, 1950, V, § 42, II, 6). This concludes the proof.

4.8. At this point it may be appropriate to introduce some classical terminology (Niemyckij and Stepanov, 1949). Given a continuous lsd system T on P and a point $x \in P$, if the +trajectory $x \mathsf{T} \mathsf{R}^+$ is contained in some quasi-compact set (i.e. the assumption in 4.7), then x or $x \mathsf{T} \mathsf{R}^+$ are called *Lagrange +stable*; if $x \in L_x^+$ then x or $x \mathsf{T} \mathsf{R}^+$ are called *Poisson +stable.*

The stability appearing here is rather a courtesy title, and it is preferred to reserve the term to describe another type of phenomena; one of these, orbital stability, is introduced below.

4.9. Definition. Let there be given a continuous lsd system T on P and a set $X \subset P$. If X has arbitrary small +invariant neighbourhoods, then it is called *orbitally +stable* (*o+ stable*); if X is $o+$ stable and, for some neighbourhood U of X, one has

$$L_x^+ \subset \bar{X} \quad \text{for all} \quad x \in U\,,$$

then X is called *asymptotically orbitally +stable* (*ao+* stable). For continuous ld systems one also defines *o−* and *ao− stability* by orientation change, and *o stability* and *ao* stability by conjunction.

More generally, if X, Y are subsets of the carrier of a continuous lsd system and $L_y^+ \subset \bar{X}$ whenever $y \in Y$, then Y is called a *region of attraction* for X; and if to any neighbourhood U of X there is a $\theta \in \mathsf{R}^+$ with

$$Y \mathsf{T} [\theta, +\infty) \subset U\,,$$

then Y will be termed a *region of uniform attraction* for X.

4.10. Remarks. The classical definition terms a subset X orbitally +stable if to any neighbourhood U of X there is a second neighbourhood V of X such that

$$x \mathsf{T} \theta \in U \quad \text{for all} \quad x \in V,\ \theta \in \mathsf{R}^+\,;$$

this latter condition is more elegantly formulated in our notation as

$$V \mathsf{T} \mathsf{R}^+ \subset U$$

whereupon $X \subset V \subset V\mathsf{T}\,\mathsf{R}^+$ and $V\mathsf{T}\,\mathsf{R}^+$ is indeed a +invariant neighbourhood of X.

Observe that an $o+$ stable set is $ao+$ stable iff some region of attraction is a neighbourhood. "Region of attraction" is to be treated as an indivisible term: it is not implied that the set considered is a region.

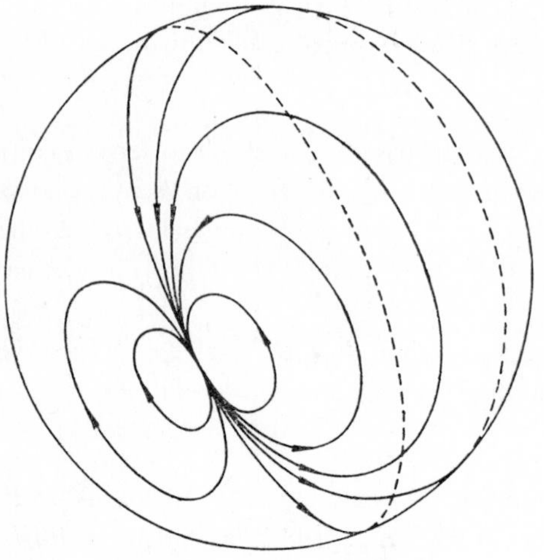

Fig. 3. Non-orbitally +stable point with region of attraction as neighbourhood.

4.11. Example. In the situation of 4.9, a set X may have a region of attraction as neighbourhood without being $o+$ stable. This may be illustrated by the differential dynamical system in the complex plane R^2 defined by $dz/d\theta = z^2$ and interpreted on the complex sphere S^2. (The point at infinity is non-critical, cf. I,3.5.) There is $L_x^+ = (0)$ for all $x \in \mathsf{S}^2$ so that S^2 is a region of attraction for 0, but 0 is obviously not $o+$ stable (cf. Fig. 3).

4.12. Lemma. If T *is a continuous lsd system on a* T_1 *space P, then every* $o+$ *stable set is +invariant.*

Proof. In a T_1 space, X coincides with the intersection of all its neighbourhoods; apply 4.9 and II,5.6.4.

The following assertions have straightforward proofs.

4.13. Lemma. Let T *be a continuous lsd system on P. Then*

1. $\emptyset$ *and P are* $o+$ *stable.*

2. If $X_i \subset P$ *are* $o+$ *stable, then so is* $\bigcup X_i$.

3. If $X_k \subset P$ *are* $o+$ *stable, then so is* $\bigcap_{k=1}^{n} X_k$ *(n, k integers).*

4. If X_1, X_2 *are separated by open sets and* $X_1 \cup X_2$ *is* $o+$ *stable, then both* X_k *are* $o+$ *stable.*

4.14. Lemma. Let T *be a continuous lsd system on P. Then for subsets of P,*

1. If Y is a region of attraction for X and $Y' \subset Y$, $X \subset X'$, *then* Y' *is a region of attraction for* X'.

2. If Y_i *are regions of attraction for X, then so is* $\bigcup Y_i$.

3. If Y is a region of attraction for closed X_i, then Y is a region of attraction for $\bigcap X_i$.

4. If Y consists of points with finite escape times and X is arbitrary, then Y is a region of attraction for X.

In particular, then, every subset $X \subset P$ has a maximal region of attraction Y, and any subset $Y' \subset P$ is a region of attraction for X iff $Y' \subset Y$.

4.15. *Lemma. Let* T *be a continuous lsd system on P. Then for subsets of P,*

1. If P is a T_1 space, then every region of uniform attraction for a set X is a region of attraction for X.

2. If Y is a region of uniform attraction for X and $Y' \subset Y$, $X \subset X'$, then Y' is a region of uniform attraction for X.

3. If Y_k are regions of uniform attraction for X, then so is $\bigcup_{k=1}^{n} Y_k$ (n, k integers).

CHAPTER V

Extension of Dynamical Systems

Assume that $\mathsf{T} \subset \mathsf{T}'$ for T an lsd system and T' an ld system, both abstract and not necessarily on the same carrier. It was proved in II,2.7 that necessarily T then has the $-$unicity property; can anything more be said concerning T? The outstanding result of the present chapter is that the answer is negative; in other words, to an arbitrary lsd system T with $-$unicity one may construct an ld system $\mathsf{T}' \supset \mathsf{T}$.

Another question of principal importance is whether, to a given non-global ld system T, there exists a gd system $\mathsf{T}' \supset \mathsf{T}$. From the second local determinacy theorem of Chapter III there is at least the negative result, that if such a global extension is at all possible, then the carrier of T' cannot coincide with that of T. It will be shown that the answer is affirmative, and a canonic maximal extension will be exhibited.

Finally, consider the following procedure. Start with a gd system, relativize over a t-open set, and then restrict and relativize again over a $+$invariant set; there results an lsd system with $-$unicity. It will be shown that this process is the most general one, in the sense that any lsd system with $-$unicity can be obtained as the restriction-*cum*-relativization of a gd system over the intersection of a t-open and a $+$invariant set.

Obviously one may pose similar, and possibly even more interesting, questions within the category of continuous dynamical systems. These are treated in Section 2, and some properties of the constructed objects are examined in Section 3. Example 3.7.1 signalizes a source of further difficulties, with a partial solution exhibited in 3.8. The exposition is based on Hájek 9, 1965.

1. EXTENSION OF ABSTRACT DYNAMICAL SYSTEMS

Throughout this section there is assumed given an lsd system T with $-$unicity on a set P (except for 1.14–16 where explicit assumptions are stated). A construction will now be performed, in separately numbered steps, and some assertions established concerning the resulting objects.

1.1. First take the trivial gd system on P and the natural gd system on R^1, and form their direct product T_0; according to II,3.2, T_0 is a gd system on $P \times \mathsf{R}^1$ also described by

$$(x, \lambda)\,\mathsf{T}_0\,\theta = (x, \lambda + \theta)$$

for all $(x, \lambda) \in P \times \mathsf{R}^1$, $\theta \in \mathsf{R}^1$.

1.2. Next, define a relation $\sim$ on $P \times \mathsf{R}^1$ by symmetrizing the relation between

$$(x, \lambda) \quad \text{and} \quad (x\,\mathsf{T}\,\varepsilon, \lambda - \varepsilon) \quad \text{with} \quad 0 \leqq \varepsilon < \alpha_x .$$

Evidently, $\sim$ is precisely the t-equivalence of II,6.1 on the direct product $(P \times \mathsf{R}^1, \mathsf{T}_1)$ of (P, T) with orientation-changed $(\mathsf{R}^1, +)$. From II,3.2, T_1 is an lsd system with $-$unicity on $P \times \mathsf{R}^1$, and then from II,6.2, $\sim$ is an equivalence relation on $P \times \mathsf{R}^1$.

Easily, $\sim$ is compatible with the gd system T_0 of 1.1: $(x, \lambda) \sim (x\,\mathsf{T}\,\varepsilon, \lambda - \varepsilon)$ implies

$$(x, \lambda)\,\mathsf{T}_0\,\theta = (x, \lambda + \theta) \sim (x\,\mathsf{T}\,\varepsilon, \lambda + \theta - \varepsilon) = (x\,\mathsf{T}\,\varepsilon, \lambda - \varepsilon)\,\mathsf{T}_0\,\theta\,;$$

and since T_0 is global, $\sim$ is strongly compatible with T_0.

1.3. In the special case that T is the semi-dynamical restriction of an ld system T' on P, $\sim$ may be described even more simply: $(x, \lambda) \sim (x', \lambda')$ iff

$$x = x'\,\mathsf{T}'\,\theta\,, \quad \lambda = \lambda' - \theta \quad \text{for some} \quad \theta \in \mathsf{R}^1 .$$

1.4. Let $\hat{\mathsf{T}}$ be the factor system of $(P \times \mathsf{R}^1, \mathsf{T}_0)$ modulo $\sim$; thus $\hat{\mathsf{T}}$ is a gd system on the set $P^\wedge = (P \times \mathsf{R}^1)/\sim$.

Also define a map

$$p_0 : P \to P \times \mathsf{R}^1\,, \quad p_0(x) = (x, 0)\,,$$

and then a map $p : P \to P^\wedge$ by composing $p_0 : P \to P \times \mathsf{R}^1$ with $e : P \times \mathsf{R}^1 \to P^\wedge$, the canonic quotient mapping corresponding to $\sim$. Thus, more directly

$$p(x) = e\,p_0(x) = e(x, 0)$$

is the equivalence class modulo $\sim$ containing $(x, 0)$.

1.5. Lemma. $p : (P, \mathsf{T}) \to (P^\wedge, \hat{\mathsf{T}})$ *is a monomorphism in* $\mathscr{D}_{slu}$; *if* T *is the lsd restriction of an ld system* T', *then* $p : (P, \mathsf{T}') \to (P^\wedge, \hat{\mathsf{T}})$ *is a monomorphism in* $\mathscr{D}_l$.

Proof. Assume $p(x_1) = p(x_2)$, i.e. $(x_1, 0) \sim (x_2, 0)$. Then one has

$$x_1 = x_2\,\mathsf{T}\,\varepsilon\,, \quad 0 = 0 - \varepsilon$$

for some $\varepsilon \geqq 0$; hence $\varepsilon = 0$; $x_1 = x_2$. Next, if $x\,\mathsf{T}\,\theta$ is defined in P, then

$$p_0(x\,\mathsf{T}\,\theta) = (x\,\mathsf{T}\,\theta, 0) \sim (x, \theta) = (x, 0)\,\mathsf{T}_0\,\theta = p_0(x)\,\mathsf{T}_0\,\theta\,;$$

since $p = ep_0$ and e is a morphism in $\mathscr{D}_g$, this yields

$$p(x \mathsf{T} \theta) = p(x) \overset{\wedge}{\mathsf{T}} \theta ,$$

as asserted. The last statement is proved analogously.

Having 1.5 we may and shall identify P with $p(P)$, thereby obtaining

$$P \subset P^{\wedge} , \quad \mathsf{T} \subset \overset{\wedge}{\mathsf{T}} ;$$

if T is the lsd restriction of an ld system T' on P, then also $\mathsf{T}' \subset \overset{\wedge}{\mathsf{T}}$. With this convention one has the following

1.6. Lemma. P generates $P^{\wedge}$.

Proof. Since the quotient map $e : P \times \mathsf{R}^1 \to P^{\wedge}$ is onto, it suffices to show that $p_0(P) = P \times (0)$ generates $P \times \mathsf{R}^1$ relative to T_0. This is immediate from $(x, \lambda) = (x, 0) \mathsf{T}_0 \lambda$.

1.7. Now set

$$P^+ = P \overset{\wedge}{\mathsf{T}} \mathsf{R}^+ , \quad P^- = P \overset{\wedge}{\mathsf{T}} \mathsf{R}^- .$$

From II,5.6, P^+ is $+$invariant and P^- $-$invariant in $P^{\wedge}$. Results concerning these two subsets of $P^{\wedge}$ are collected in 1.8–11.

1.8. Lemma. P^+ *is* $+$*invariant in* $P^{\wedge}$, *and the relativization of* $\overset{\wedge}{\mathsf{T}}$ *to* P^+ *is a gsd system* T^+ *on* P^+; *furthermore,* $\mathsf{T} \subset \mathsf{T}^+$, T^+ *has* $-$*unicity and the same s-points as* T, *P semi-generates* P^+.

Proof. Existence of $\mathsf{T}^+ \subset \overset{\wedge}{\mathsf{T}}$ on P^+, globality and $-$unicity all follow from II,5.4. Obviously $\mathsf{T} \subset \mathsf{T}^+$, P semi-generates P^+:

$$P^+ = P \overset{\wedge}{\mathsf{T}} \mathsf{R}^+ = P \mathsf{T}^+ \mathsf{R}^+ .$$

It remains to prove the assertion on s-points; let S and S^+ be the corresponding s-point sets. From $\mathsf{T} \subset \mathsf{T}^+$ (or II,3.15) one has $P \cap S^+ \subset S$, and from II,5.13, $S^+ \subset P$; thus

$$S^+ \subset S . \tag{1}$$

Now take, in P, any non-s-point of P^+; working in $P \times \mathsf{R}^+$, one has

$$(x, 0) \sim (x', \lambda') \mathsf{T}_0 \theta , \quad \lambda' \geqq 0 < \theta .$$

Since there cannot be $\lambda' + \theta = -\varepsilon$ with $\varepsilon \geqq 0$, it follows that

$$x = x' \mathsf{T} \varepsilon , \quad 0 = \lambda' + \theta - \varepsilon ,$$

so that $\varepsilon = \lambda' + \theta > 0$ and x is not an s-point of P. This proves

$$P - S^+ \subset P - S$$

and, with (1), yields $S = S^+$. The proof of 1.8 is thus concluded.

1.9. Lemma. P^- *is t-open and* $-$*invariant in* $P^\wedge$, *and the relativization of* $\overset{\wedge}{\mathrm{T}}$ *to* P^- *is an ld system* T^- *on* P^-; *furthermore,* $\mathrm{T} \subset \mathrm{T}^-$, *and* P *generates* P^-. *If* T *is the lsd restriction of an ld system* T' *on* P, *then also* $\mathrm{T}' \subset \mathrm{T}^-$, *and* P *is t-open in* P^- *(and hence also in* $P^\wedge$*).*

Proof. Obviously P^- is the image of $P \times \mathrm{R}^-$ under e. Take any $(x, \lambda) \in P \times \mathrm{R}^-$. If $\theta \leqq 0$, then

$$(x, \lambda)\, \mathrm{T}_0\, \theta = (x, \lambda + \theta) \in P \times \mathrm{R}^- ;$$

if $0 \leqq \theta < \alpha_x$, then

$$(x, \lambda)\, \mathrm{T}_0\, \theta = (x, \lambda + \theta) \sim (x\, \mathrm{T}\, \theta, \lambda) \in P \times \mathrm{R}^- .$$

Hence $e(x, \lambda)\, \overset{\wedge}{\mathrm{T}}\, \theta \in P^-$ for all $e(x, \lambda) \in P^-$, $\theta < \alpha_x$; this proves that P^- is t-open $-$invariant in $P^\wedge$. Then existence of the ld system $\mathrm{T}^- \supset \mathrm{T}$ on P^- follows from III,1.13. Obviously P generates P^-:

$$P^- = P\, \overset{\wedge}{\mathrm{T}}\, \mathrm{R}^- = P\, \mathrm{T}^-\, \mathrm{R}^- \subset P\, \mathrm{T}^-\, \mathrm{R}^1 \subset P^- .$$

Finally, if T is the lsd restriction of an ld system T' on P, then $\mathrm{T}' \subset \mathrm{T}^-$ follows from 1.5, and t-openness of P in P^- from III,1.13 again.

1.10. Lemma. $P^- \cap P^+ = P$, $P^- \cup P^+ = P^\wedge$.

Proof. The second formula follows directly from 1.7:

$$P^- \cup P^+ = (P\, \overset{\wedge}{\mathrm{T}}\, \mathrm{R}^-) \cup (P\, \overset{\wedge}{\mathrm{T}}\, \mathrm{R}^+) = P\, \overset{\wedge}{\mathrm{T}}\, (\mathrm{R}^- \cup \mathrm{R}^+) = P\, \overset{\wedge}{\mathrm{T}}\, \mathrm{R}^1 = P^\wedge .$$

Now consider the first; obviously

$$P \subset P^- \cap P^+ . \tag{2}$$

To obtain the opposite inclusion, work in $P \times \mathrm{R}^1$, and assume that

$$(x, \lambda) \sim (x', \lambda'), \quad \lambda \geqq 0 \geqq \lambda' .$$

Then, for some $\varepsilon \geqq 0$, either

$$x = x'\, \mathrm{T}\, \varepsilon, \quad \lambda = \lambda' - \varepsilon ,$$

implying $\lambda = 0$ and $(x, \lambda) = (x, 0) \in p_0(P)$; or

$$x' = x\, \mathrm{T}\, \varepsilon, \quad \lambda' = \lambda - \varepsilon ,$$

yielding $\varepsilon = \lambda - \lambda' \geqq \lambda \geqq 0$. Since $x\, \mathrm{T}\, \varepsilon$ is defined, so is $x\, \mathrm{T}\, \lambda$, and then

$$(x, \lambda) \sim (x\, \mathrm{T}\, \lambda, 0) \in p_0(P) .$$

Thus in both cases

$$e(x, \lambda) = e(x', \lambda') \in e\, p_0(P) = P .$$

This yields the inclusion opposite to (2) and hence completes the proof of 1.10.

1.11. Corollary. P is $+$invariant in P^- (so that $\alpha_x = \alpha_x^-$ for $x \in P$), $-$invariant in P^+, and t-convex in $P^\wedge$. Each critical or feebly critical point of $P^\wedge$ and each cycle of $P^\wedge$ is contained within P.

Proof. For the first assertion use $P = P^- \cap P^+$, 1.7, II,5.4 and II,5.10. For the second, assume that $x \hat{\mathsf{T}} \theta = x$ with $x \in P^\wedge$, $\theta > 0$. Since P generates $P^\wedge$ (cf. 1.6), $x \hat{\mathsf{T}} \mathsf{R}^1$ must intersect P, say at a point y; from II,4.3 $y \hat{\mathsf{T}} \theta = y \in P$; from II,4.9 and t-convexity

$$x \in x \hat{\mathsf{T}} \mathsf{R}^1 = y \hat{\mathsf{T}} \mathsf{R}^1 = y \hat{\mathsf{T}} [0, \theta] \subset P \,,$$

concluding the proof.

The results thus obtained will now be summarized and reorganized. The statements on $-$unicity are obtained from II,5.14–19; in applying these latter, the requirements concerning s-points and t-convexity follow from 1.8 and 1.11.

1.12. Theorem. Any lsd system with $-$unicity (P, T) may be represented as the relativization of a gsd system (P^+, T^+) to its $-$invariant subset; the system (P^+, T^+) has $-$unicity, the same s-points as (P, T), and P^+ is semi-generated by P. The gsd extension (P^+, T^+) of (P, T) is determined uniquely by these conditions, up to isomorphism in $\mathscr{D}_{sgu}$.

Proof. 1.8, 1.11, II,5.15.

The extension procedure from (P, T) to (P^+, T^+) may be described as the action of a covariant functor $E_{12} : \mathscr{D}_{slu} \to \mathscr{D}_{sgu}$; to any morphism $f : (P_1, \mathsf{T}_1) \to (P_2, \mathsf{T}_2)$ in $\mathscr{D}_{slu}$, E_{12} assigns the morphism $f^+ : (P_1^+, \mathsf{T}_1^+) \to (P_2^+, \mathsf{T}_2^+)$ in $\mathscr{D}_{sgu}$, defined by exploiting commutativity in the diagram

$$\begin{array}{ccccc} P_1 & \to & P_1 \times \mathsf{R}^+ & \to & P_1^+ \\ {\scriptstyle f}\downarrow & & \downarrow & & \downarrow{\scriptstyle f^+} \\ P_2 & \to & P_2 \times \mathsf{R}^+ & \to & P_2^+ \end{array}$$

noting that $(x, \lambda) \sim (x \mathsf{T}_1 \varepsilon, \lambda - \varepsilon)$ implies

$$(f(x \mathsf{T}_1 \varepsilon), \lambda - \varepsilon) = (f(x) \mathsf{T}_2 \varepsilon, \lambda - \varepsilon) \sim (f(x), \lambda) \,.$$

Similar categorial interpretations may be given to the following statements (also see 1.18).

1.13. Proposition. Any lsd system with $-$unicity (P, T) may be represented as the relativization of an ld system (P^-, T^-) to its $+$invariant subset; P^- is then generated by P.

Proof. 1.9, 1.11.

1.14. Proposition. Any ld system (P, T) *may be represented as the relativization of an ld system* (P^-, T^-) *to its t-open +invariant subset, with* P^- *generated by P and*

$$\alpha_x^- = \alpha_x\,, \quad \beta_x^- = -\infty \quad \text{for} \quad x \in P\,.$$

Proof. 1.9, 1.11.

1.15. Proposition. Any ld system (P, T) *may be represented as the relativization of an ld system* $(P^\vee, \mathsf{T}^\vee)$ *to its t-open −invariant subset, with* $P^\vee$ *generated by P and*

$$\alpha_x^\vee = +\infty\,, \quad \beta_x^\vee = \beta_x \quad \text{for} \quad x \in P\,.$$

Proof. 1.14 preceded and followed by orientation change.

1.16. Theorem. Any ld system (P, T) *may be represented as the relativization of a gd system* $(P^\wedge, \hat{\mathsf{T}})$ *to its t-convex subset;* $P^\wedge$ *is then generated by P. The gd extension* $(P^\wedge, \hat{\mathsf{T}})$ *of* (P, T) *is determined uniquely by these conditions, up to isomorphism in* $\mathscr{D}_g$.

Proof. 1.5, 1.6, 1.9, II,5.17.

1.17. Theorem. Any lsd system with −unicity (P, T) *may be represented as the relativization of a gd system* $(P^\wedge, \hat{\mathsf{T}})$ *to the intersection of a +invariant and a −invariant subset, and* $P^\wedge$ *is then generated by P. The gd extension* $(P^\wedge, \hat{\mathsf{T}})$ *is determined uniquely by these conditions, up to isomorphism in* $\mathscr{D}_g$.

Proof. 1.4−6, 1.9−10, II,5.19.

1.18. The constructions yielding 1.12−17 are closely related; a more adequate formulation is that commutativity holds in the following functor diagram

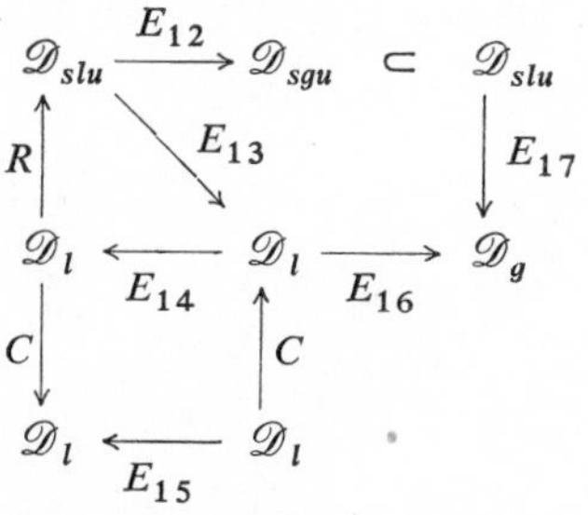

Here E_{ij} is the extension functor described in item 1.*ij* (for example E_{12} in 1.12), C the orientation changing functor, and R the lsd restriction functor, II,3.10−12.

1.19. In the situation of 1.12, proposition II,5.14 also yields that, to any gsd extension $(P', \mathsf{T}') \supset (P, \mathsf{T})$, there is a morphism

$$f : (P^+, \mathsf{T}^+) \to (P', \mathsf{T}') \quad \text{in} \quad \mathscr{D}_{sg}\,,$$

uniquely determined by $f(x) = x$ for $x \in P$ (obviously f maps onto the subset generated by P in P'). In this sense, (P^+, T^+) is the *maximal gsd extension* of (P, T). Similar remarks apply to 1.16 and 1.17, referring to II,5.16 and II, 5.18 respectively.

2. EXTENSION OF CONTINUOUS DYNAMICAL SYSTEMS

The construction of the preceding section will now be followed through for the case of continuous dynamical systems. Explicitly, the assumptions are that there is given a continuous lsd system T with $-$unicity on a topological space P, with topology τ.

In the standard manner, the set $P \times \mathsf{R}^1$ is endowed with the product topology of τ with the natural topology of R^1; on $P^\wedge = (P \times \mathsf{R}^1) \sim$ one then has the quotient topology $\hat{\tau}$, and on P^+ and P^- the subset topologies τ^+ and τ^- inherited from $P^\wedge$. The notation of Section 1 will be used without further reference.

2.1. Lemma. The maps p_0, e, p are continuous; T_0, T_1, $\hat{\mathsf{T}}$, T^+, T^- *are continuous dynamical systems*; P^- *is open in* $P^\wedge$, *and* P *is open in* P^+.

Proof. 2.1.1. The assertions concerning T_0, T_1 follow from IV,3.4; that concerning $\hat{\mathsf{T}}$, e from IV,3.8 (recall that $\sim$ is strongly compatible with T_0 since T_0 is global, see 1.2); as for T^+, apply IV,2.10 to P^+ $+$invariant in $P^\wedge$; p_0 is obviously continuous, and hence so is $p = ep_0$. The assertion on T^- will follow from IV,2.9, and that on openness of P in P^+ from 1.10, once it is proved that P^- is open in $P^\wedge$.

2.1.2. Consider the set $e^{-1}(P^-) = e^{-1}(P \,\hat{\mathsf{T}}\, \mathsf{R}^-)$; by definition, it consists of all points in $P \times \mathsf{R}^1$ $\sim$equivalent some point of $P \times \mathsf{R}^-$. Hence

$$e^{-1}(P^-) = \{(x, \lambda) : x \in P, \lambda < \alpha_x\}. \tag{1}$$

Indeed, if

$$(x, \lambda) \sim (x', \lambda') \in P \times \mathsf{R}^-,$$

then $\lambda \leqq 0$ and, for some $\varepsilon \geqq 0$, either

$$x = x' \mathsf{T} \varepsilon, \quad \lambda = \lambda' - \varepsilon \leqq -\varepsilon < \alpha_x$$

or

$$x' = x \mathsf{T} \varepsilon, \quad \lambda' = \lambda - \varepsilon,$$

implying

$$\lambda = \lambda' + \varepsilon \leqq \varepsilon < \alpha_{x'} + \varepsilon = \alpha_x - \varepsilon + \varepsilon = \alpha_x.$$

Conversely, if an (x, λ) has $\lambda < \alpha_x$, then either $\lambda \leqq 0$, and directly

$$(x, \lambda) \in P \times \mathsf{R}^-;$$

or $0 < \lambda < \alpha_x$ and

$$(x, \lambda) \sim (x \mathsf{T} \lambda, 0) \in P \times \mathsf{R}^- .$$

This proves (1). Since escape times are lower semi-continuous, IV,1.4, the set described in (1) is open in $P \times \mathsf{R}^1$; hence and by definition of the quotient topology, P^- is open in $P^\wedge$. This concludes the proof of 2.1.

2.2. *Lemma. P is open in $P^\wedge$ iff T has the openness property. In the positive case the relation $\sim$ is open and the map $p : P \to P^\wedge$ is interior.*

Proof. 2.2.1. Since by assumption T is a continuous lsd system with $-$unicity, T has openness iff it is the lsd restriction of a continuous ld system T' on P (cf. IV,2.5).

If such a T' exists, then it is easily shown using 1.3 (a similar proof appears in 2.1) that

$$e^{-1}(P) = e^{-1}\, e(P \times (0)) = \mathit{domain}\ \mathsf{T}' ,$$

a set open in $P \times \mathsf{R}^1$ by definition IV,1.1.2; hence P is open in $P^\wedge$.

2.2.2. In any case $\hat{\mathsf{T}}$ and its semi-dynamical restriction do have the openness property (2.1 and IV,2.3). If P is open in $P^\wedge$, then T, the relativization of $\hat{\mathsf{T}}$ to P, has the openness property according to IV,2.10. This completes the proof of the first assertion.

2.2.3. Now assume that T' exists. From 1.2–3 and II,6.3, the least $\sim$saturated set containing a given open set G in $P \times \mathsf{R}^1$ is

$$H = G\, \mathsf{T}_1'\, \mathsf{R}^1 , \tag{2}$$

where T_1' is the continuous ld system on $P \times \mathsf{R}^1$ with lsd restriction T_1; obviously T_1' exists iff T_1 has the openness property (cf. IV,2.5, IV,3.4; however, T_1' may be described directly using T'). Hence from IV,2.3, H is open and $\sim$ is then an open equivalence relation by definition.

If P is open in $P^\wedge$ then, obviously, the injection $p : P \to P^\wedge$ of 1.5 is an interior mapping. This concludes the proof of 2.2.

For continuous dynamical systems one then has statements parallel to 1.12–18, except that unicity of the extension is often not asserted; the reason for this is that in the continuous case IV,2.15 replaces II,5.14.

2.3. *Proposition. Any continuous lsd system with $-$unicity (P, T, τ) may be represented as the relativization of a continuous gsd system $(P^+, \mathsf{T}^+, \tau^+)$ to its open $-$invariant subset. Furthermore, $(P^+, \mathsf{T}^+, \tau^+)$ has $-$unicity, the same s--points as T and P^+ is semi-generated by P.*

Proof. 1.12, 2.1.

2.4. Proposition. Any continuous lsd system (P, T, τ) *with* $-$*unicity and openness may be represented as the relativization of a continuous ld system* $(P^-, \mathsf{T}^-, \tau^-)$ *to its* $+$*invariant subset*; P^- *is then generated by* P.

Proof. 1.13, 2.1.

2.5. Theorem. Any continuous ld system (P, T, τ) *may be represented as the relativization of a continuous gd system* $(P^\wedge, \hat{\mathsf{T}}, \hat{\tau})$ *to its open t-convex subset* P *which generates* $P^\wedge$. *These conditions determine* $(P^\wedge, \hat{\mathsf{T}}, \hat{\tau})$ *uniquely up to isomorphism in* $\mathscr{D}_g^T$.

Proof. 1.16, 2.1–2, IV,2.15. As in 1.19, $(P^\wedge, \hat{\mathsf{T}}, \hat{\tau})$ will be termed the *maximal gd extension* of (P, T, τ).

2.6. Proposition. Any continuous lsd system (P, T, τ) *with* $-$*unicity and openness may be represented as the relativization of a continuous gd system* $(P^\wedge, \hat{\mathsf{T}}, \hat{\tau})$ *to the intersection of a* $+$*invariant subset with an open* $-$*invariant subset*; $P^\wedge$ *is generated by* P.

Proof. 1.17, 2.1–2.

To appreciate the rôle of the openness requirement in 2.4 and 2.6, consider again the situation of Theorem 1.17, and let T be compatible with a topology τ on P. Then $\hat{\mathsf{T}}$ is also compatible with the topology $\hat{\tau}$ on $P^\wedge$ (cf. 2.1), and one has $P \subset P^\wedge$. However, the given topology τ on P need not coincide with the subset topology iherited from $P^\wedge$; they do coincide if T has the openness property, but in general one merely has that the inclusion map $p : P \to P^\wedge$ is continuous and 1–1.

The preceding results may obviously be employed as proof apparatus in assertions in which the constructions of $\hat{\mathsf{T}}$, etc., do not intervene directly.

2.7. Lemma. Let T *be a continuous lsd system with* $-$*unicity and openness on* P. *Then*

$$x_i \mathsf{T} \theta_i \to x \mathsf{T} \theta \text{ in } P \quad \text{and} \quad \theta_i \to \theta \text{ in } \mathsf{R}^+ \quad \text{implies} \quad x_i \to x .$$

Proof. From 2.4, $x_i = (x_i \mathsf{T} \theta_i) \mathsf{T}^- (-\theta_i) \to (x \mathsf{T} \theta) \mathsf{T}^- (-\theta) = x$ in P.

2.8. Lemma. If T *is a continuous lsd system with* $-$*unicity and openness on* P, *then each translation* T_θ *is a homeomorphism with domain* T_θ *open in* P; *if, furthermore,* T *is global, then* $T_\theta : P \approx P \mathsf{T} \theta$.

Proof. IV,1.14, 2.7.

Finally, two results applying to ld systems will be extended to lsd systems with $-$unicity, namely IV,2.16–17 and IV,4.6.

2.9. Lemma. Let T *be a continuous lsd system with* $-$*unicity and openness on* P, $Q \subset P$ *a closed section with t-extent* $\varepsilon > 0$, *and assume that* $I = [0, \varepsilon]$ *has* $Q \times I \subset$ *domain* T. *Then* T *partializes to a homeomorphism* $\mathsf{T}_p = \mathsf{T} \mid (Q \times I)$,

$$\mathsf{T}_p : Q \times I \approx Q \mathsf{T} I .$$

Hence the composition of T_p^{-1} *with the projection* $Q \times I \to Q$ *is a retracting deformation* $r : Q \mathsf{T} I \to Q$ *with*

$$r(x \mathsf{T} \theta) = x \quad \textit{for} \quad x \in Q,\ 0 \leqq \theta \leqq \varepsilon .$$

Proof. As in IV,2.16, applying II,6.7.3 and 2.7.

2.10. Lemma. For continuous lsd systems with $-$*unicity and openness, each* $+$*limit set is closed invariant.*

Proof. Having IV,4.4, it is only required to show that a $+$limit set L_x^+ is $-$invariant, i.e. that its complement is $+$invariant. Assume the contrary; then there exists $x' \notin L_x^+$, $\theta \in \mathsf{R}^+$ with $x' \mathsf{T} \theta \in L_x^+$, i.e. with

$$x \mathsf{T} \theta_n \to x' \mathsf{T} \theta$$

for some $\theta_n \to +\infty$ in R^+. But then

$$\big(x \mathsf{T} (\theta_n - \theta)\big) \mathsf{T} \theta = x \mathsf{T} \theta_n \to x' \mathsf{T} \theta ,$$

and 2.7 yields

$$x \mathsf{T} (\theta_n - \theta) \to x' ,$$

i.e. $x' \in L_x^+$; this contradiction proves the assertion.

2.11. The assertions on openness in 2.3 and 2.5 should be emphazised; they state that a continuous dynamical system with $-$unicity behaves, locally at each point, as if it were global. Further consequences of this property appear in the following section.

Since in classical differential equation theory there seems to be nothing even remotely reminiscent of the preceding extension procedure, an illustrative example would be well in place at this point.

2.12. Example. Consider again II,1.3.2, i.e. the autonomous differential equation in R^1,

$$\frac{dx}{d\theta} = x^2 + 1 ; \qquad (3)$$

the corresponding continuous ld system is described by

$$x \mathsf{T} \theta = \tan\big(\theta + \arctan x\big) .$$

Now follow through construction 1.1–4: $P \times \mathsf{R}^1$ is euclidean

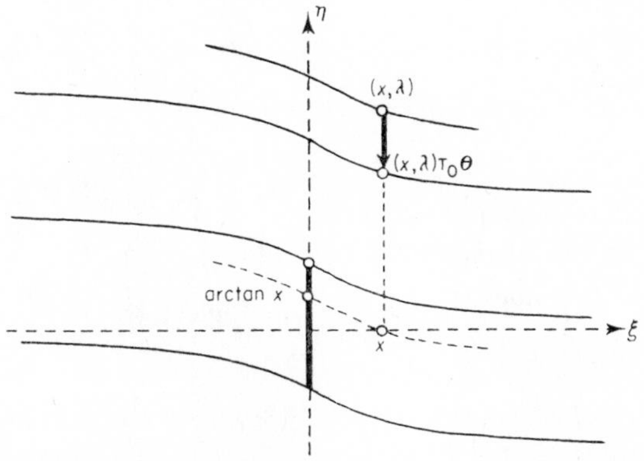

Fig. 1. Equivalence classes modulo $\sim$ in gd extension of (3).

2-space; on introducting orthogonal coordinates (ξ, η), the equivalence classes modulo $\sim$ are curves, with equations

$$\eta = (\arctan x + \lambda) - \arctan \xi$$

for the equivalence class containing the point (x, λ) (see Fig. 1). The gd system T_0 which induces $\hat{\mathsf{T}}$ is described by uniform motion in the direction of the η-axis (i.e. upward in Fig. 1). There are obvious sections for the relation $\sim$, and the η-axis itself is a reasonable choice. On this axis, the original carrier R^1 is represented on the open interval $(-\frac{1}{2}\pi, \frac{1}{2}\pi)$ by means of the assignment $x \to \arctan x$, and the original ld system by the relativization of the natural gd system on the η-axis.

2.13. The preceding example suggests an interesting interpretation of part of the maximal gd system construction: in 1.2, for $0 \leqq \varepsilon < \alpha_x$ and $\theta = \varepsilon - \lambda$ one has

$$(x, \lambda) \sim (x \mathsf{T} \varepsilon, \lambda - \varepsilon) = (t_x(\varepsilon), \lambda - \varepsilon) = (t_x(\theta + \lambda), -\theta) .$$

Thus, equivalence classes modulo $\sim$ are, essentially, graphs of solutions and of solutions shifted along the θ-axis (up to change of orientation and interchange of "axes").

3. OPENNESS CONDITIONS

The heading concerns these two assertions: the carrier P of a continuous ld system T is open in the carrier $P^\wedge$ of its maximal gd extension, and the extended gd system $\hat{\mathsf{T}}$ has the openness property (cf. IV, 2.3). These are the background of the following fundamental lemma (the notation of Sections 1 and 2 is preserved).

3.1. Lemma. If T *is a continuous ld system on P, then*

$$P^\wedge = P \hat{\mathsf{T}} \mathsf{R}^1 = \bigcup_{\theta \in \mathsf{R}^1} (P \hat{\mathsf{T}} \theta)$$

with each summand $P \hat{\mathsf{T}} \theta$ *open in* $P^\wedge$ *and homeomorphic to P,*

$$T_\theta^\wedge \mid P : P \approx P \hat{\mathsf{T}} \theta .$$

Proof. 2.5, 2.8 and IV,2.3 applied to $\hat{\mathsf{T}}$.

It follows that local topological properties are automatically inherited by $P^\wedge$. In particular,

3.2. Lemma. Let T *be a continuous ld system on P. If (and only if) P belongs to some of the classes:* T_0, T_1, *locally (pathwise) connected, locally quasi-compact, then so does* $P^\wedge$. *Assuming that* $P^\wedge$ *is* T_2, *if (and only if) P belongs to some of the classes:* T_ϱ, *LC, n-dimensional, then so does* $P^\wedge$.

Naturally, one then inquires whether some non-local properties of P also carry over to $P^\wedge$.

3.3. Proposition. Let T *be a continuous ld system on P. Then* $P^\wedge$ *is quasi-compact if P is quasi-compact, and in the positive case* $P = P^\wedge$.

Proof. In any case, from unicity in 2.5 it follows that $P = P^\wedge$ if T is global. If P is quasi-compact, then T is global by IV,1.10, whereupon $P^\wedge = P$ is indeed quasi-compact.

Conversely, assume that $P^\wedge$ is quasi-compact; then (1) describes an open cover of $P^\wedge$, so that there exist $\theta_1, \theta_2, \ldots, \theta_n$ in R^1 with

$$P^\wedge = \bigcup_1^n (P \overset{\wedge}{\mathsf{T}} \theta_k).$$

Set $\theta_0 = 1 + \max \theta_k$; then

$$P^\wedge = P^\wedge \overset{\wedge}{\mathsf{T}}(-\theta_0) = \bigcup_1^n (P \overset{\wedge}{\mathsf{T}}(-\varepsilon_k)), \quad \varepsilon_k = \theta_0 - \theta_k \geqq 1. \tag{2}$$

To prove that T is global assume the contrary, for example that $\alpha_y < +\infty$ for some $y \in P$. Then there also exists an $x \in P$ with $\alpha_x < 1$, for instance $x = y\,\mathsf{T}\,(\alpha_y - \frac{1}{2})$ (if $\alpha_y \geqq 1$; otherwise take $x = y$). From (2), there then exist $z \in P$, ε_k such that, in $P \times \mathsf{R}^1$,

$$(x, 0) \sim (z, 0)\,\mathsf{T}_0\,(-\varepsilon_k) = (z, -\varepsilon_k).$$

This yields $z = x\,\mathsf{T}\,\varepsilon_k$, implying

$$1 \leqq \varepsilon_k < \alpha_x,$$

in contradiction with $\alpha_x < 1$. Thus T is global, so that $P = P^\wedge$ is indeed quasi-compact. This concludes the proof of 3.3.

Observe that this also proves that $P^\wedge$ is compact iff it is a T_2 space and P is compact.

3.4. Lemma. Let T *be a continuous ld system on P; then* $P^\wedge$ *is connected iff P is connected.*

Proof. If P is connected, then from IV,2.8.1, $P^\wedge = P \overset{\wedge}{\mathsf{T}} \mathsf{R}^1$ is connected. Next, assume P is not connected, so that there is a non-trivial decomposition

$$P = P_1 \cup P_2$$

into disjoint open sets. Easily, both P_k are invariant; hence T relativizes on P_k to a continuous ld system T_k with the same bounds as T (IV, 2.9 and II,5.5), and obviously (P, T) is the topological direct sum of the (P_k, T_k) (cf. IV,3.6). Now let (P', T') be the direct sum of the corresponding $(P_k^\wedge, \overset{\wedge}{\mathsf{T}}_k)$; then from unicity in 2.5,

$$P_1^\wedge \cup P_2^\wedge = P' \approx P^\wedge$$

with $P_k^\wedge$ disjoint open and non-trivial in P'. Hence $P^\wedge$ is not connected, completing the proof of 3.4.

3.5. Corollary. Let T *be a continuous ld system on* P*, and assume that* $P^\wedge$ *is* T_2*; then* $P^\wedge$ *is an n-manifold iff* P *is an n-manifold.*

Proof. An n-manifold is a connected T_2 space each point of which has a neighbourhood homeomorphic to R^n; the assertions then follow from 3.4 and 3.1.

3.6. Corollary. Let T *be a continuous ld system on* P*, and assume that* $P^\wedge$ *is* T_2*; then* $P^\wedge$ *is an n-manifold with boundary iff* P *is an n-manifold with boundary, and in the positive case the boundary of* $P^\wedge$ *is generated by that of* P*.*

Proof. As in 3.5, and use the fact that any $x \in P^\wedge$ has $x = T_\theta^\wedge(y)$ with $y \in P$ and $T_\theta^\wedge \mid P$ a homeomorphism between open subsets of $P^\wedge$.

The additional assumption in 3.2 and 3.5–6 that $P^\wedge$ be a T_2 space should not be taken lightly; even if T is a differential ld system on a 2-manifold P, the carrier $P^\wedge$ of the extension need not be a T_2 space, as shown in the following example. A sufficient condition is exhibited in 3.8.

3.7.1. Example. Consider again the continuous gd system on R^2,

$$(x, y)\,\mathsf{T}'\,\theta = (x + \theta, y)\,,$$

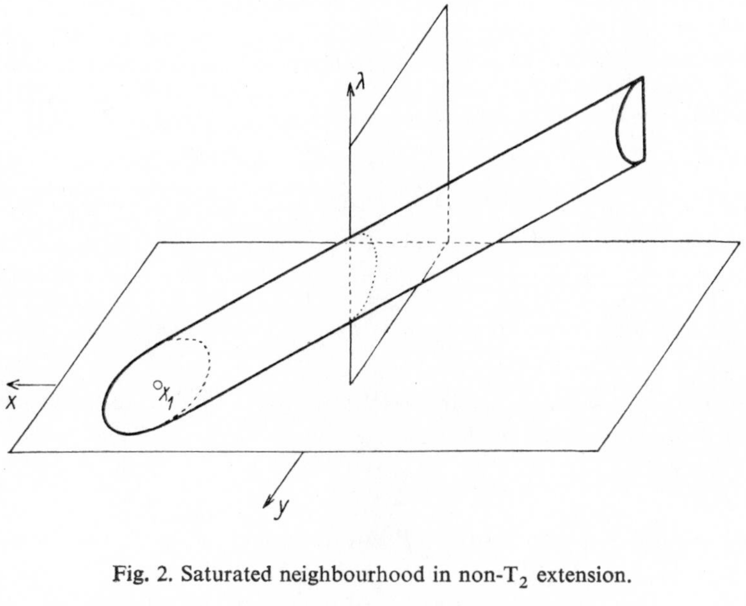

Fig. 2. Saturated neighbourhood in non-T_2 extension.

relativized to a continuous ld system T on the open subset

$$P = \{(x, y) \in \mathsf{R}^2 : x \neq 0 \text{ or } y > 0\}\,.$$

Evidently P is a 2-submanifold of R^2, and T is a differential ld system on P. Now form $P \times R^1$, T_0, $\sim$ as described in 1.1–3, and consider, for example, the equivalence classes (i.e. elements of $P^\wedge$) containing the following points of $P \times R^1$:

$$x_{-1} = ((-1, 0), 2)\,, \quad x_1 = ((1, 0), 0)\,;$$

obviously x_{-1} non $\sim x_2$. It is easily verified that $P^\wedge$ is a T_1 space as it should be according to 3.2; however, it is not T_2, since each pair of neighbourhoods of $e(x_{-1})$, $e(x_1)$ intersect. To see this, observe that in $P \times R^1$, each pair of $\sim$saturated open neighbourhoods of x_{-1}, x_1 intersect (such a neighbourhood of x_1 is indicated in Fig. 2.).

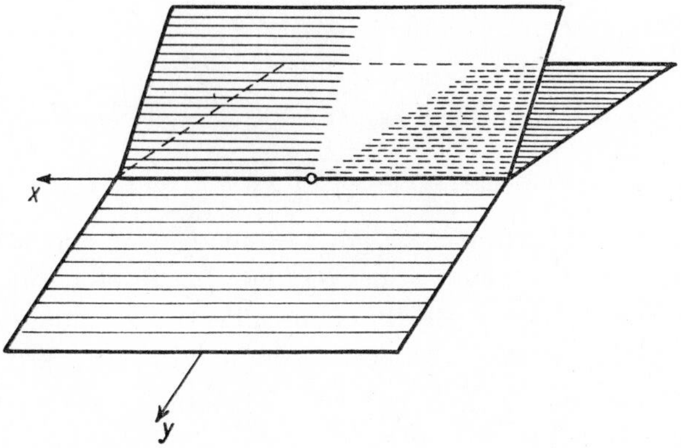

Fig. 3. Maximal gd extension: original carrier marked by dashes.

3.7.2. Example. Modify the preceding example merely by adding the origin (0, 0) to P; the resulting set is no longer a manifold with boundary. The carrier of the maximal gd extension is sketched in Fig. 3; observe that it is a T_2 space.

3.8. Proposition. Let T *be a continuous ld system on a* T_2 *space* P. *Then* $P^\wedge$ *is* T_2 *if either of the following conditions obtain:*

1. *The positive espace times* α_x *define a continuous map* $\alpha : P \to R^\#$,
2. *For each* $\theta \in R^+$,

$$x_i \to x,\, x_i T \theta \to y \text{ in } P \quad \text{imply} \quad x_i T \theta \to x T \theta\,.$$

Proof. 3.8.1. First show that 1 implies 2; thus assume 1 and the premise of 2. Then $\theta < \alpha_{x_i}$, so that $\theta \leqq \lim \alpha_{x_i} = \alpha_x$. Now $\theta = \alpha_x$ would contradict $x_i T \theta \to y$ and IV,1.11; thus $0 \leqq \theta < \alpha_x$, $x T \theta$ is defined and

$$x_i T \theta \to x T \theta$$

follows from IV, 1.1.1. This proves 2.

3.8.2. Now assume 2 and show that it implies $P^\wedge$ is T_2. Working in $P \times R^1$, take

$$(x_1, \lambda_1) \text{ non} \sim (x_2, \lambda_2)\,; \tag{3}$$

it is required to prove that the (x_k, λ_k) possess disjoint open $\sim$saturated neighbourhoods. Since $\sim$ is open (2.2 and IV,2.3), it suffices to show that

$$(y_i, \lambda_i) \to (x_1, \lambda_1), \quad (y_i \,\mathsf{T}\, \varepsilon_i, \lambda_i - \varepsilon_i) \to (x_2, \lambda_2) \tag{4}$$

in $P \times \mathsf{R}^1$ with $\varepsilon_i \in \mathsf{R}^+$ leads to a contradiction. From (4),

$$\varepsilon_i \to \lambda_1 - \lambda_2, \quad y_i \to x_1,$$

so that 2 yields $y_i \,\mathsf{T}\, \varepsilon_i \to x_1 \,\mathsf{T}\, (\lambda_1 - \lambda_2)$; since also $y_i \,\mathsf{T}\, \varepsilon_i \to x_2$ and P is T_2,

$$x_1 \,\mathsf{T}\, (\lambda_1 - \lambda_2) = x_2$$

and

$$(x_1, \lambda_1) \sim (x_1 \,\mathsf{T}\, (\lambda_1 - \lambda_2), \lambda_1 - (\lambda_1 - \lambda_2)) = (x_2, \lambda_2),$$

contradicting (3). This completes the proof of 3.8.

3.9. Remarks. According to 2.5, the gd extension $(P^\wedge, \hat{\mathsf{T}})$ may also be obtained, up to an isomorphism in $\mathscr{D}_g^T$, by pre- and post-application of an orientation change in the construction of Section 1. Hence a third sufficient condition in 3.8 is that the negative escape times $\beta_x : P \to \mathsf{R}^\#$ depend continuously on $x \in P$. In particular, the conclusion of 3.8 holds if $\beta_x \equiv -\infty$ (for example in example IV,1.12; observe that 3.8.2 is satisfied but 1 is not).

As has been seen, the carrier of the maximal gd extension may well not be a T_2 space even if the original carrier is such. Naturally the question arises whether, to a given continuous dynamical system on a T_2 space, there exists any global extension on a T_2 carrier. A necessary condition, which might be termed the *Hausdorff property*, will now be given.

3.10. Lemma. Let T *be a continuous lsd system on P, and assume that there exists a continuous gsd extension* $\mathsf{T}' \supset \mathsf{T}$ *on a* T_2 *space. Then* T *has the following property: For any* $x \in P$ *and* $\theta \in \mathsf{R}^+$ *the conditions*

$$x_i \to x \leftarrow x_i',$$

$$x_i \,\mathsf{T}\, \theta \to y, \quad x_i' \,\mathsf{T}\, \theta \to y'$$

imply $y = y'$.

Proof. $y \leftarrow x_i \,\mathsf{T}\, \theta = x_i \,\mathsf{T}'\, \theta \to x \,\mathsf{T}'\, \theta$ yields $x \,\mathsf{T}'\, \theta = y$, and $x \,\mathsf{T}'\, \theta = y'$ for a like reason.

3.11.1. Example. The continuous ld system of 3.7 does have the Hausdorff property; indeed, it may be extended to a gd system on R^2 in an obvious manner.

3.11.2. Example. In conclusion there will be exhibited a continuous ld system on a T_2 carrier which does not have the Hausdorff property. The carrier P is a subspace of R^2 and consists of all complex $z \in \mathsf{R}^2$ with $|\mathrm{Arg}\, z| \leqq \frac{2}{3}\pi$; this includes the bounding open-rays but, of course, excludes the origin. Now let τ be the ld system defined by movement along the indicated curves, with the negative real part as parameter. Obviously τ does not have the Hausdorff property (see Fig. 4).

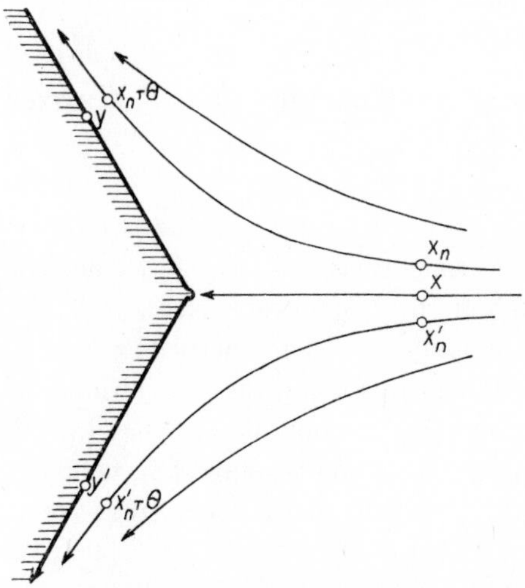

Fig. 4. Dynamical system without the Hausdorff property.

It may be remarked that this system is differential. To see this proceed as follows. First take the gd system defined by $dz/d\theta = -1$ relativized to exclude the semi-axis of non-positive reals. Then map, using the correspondence $w = z^{2/3}$; there results

$$\frac{dz}{d\theta} = -\frac{2}{3}\frac{1}{\sqrt{z}}.$$

Finally re-parametrize to obtain as new parameter $-x$ in $z = x + iy$:

$$\frac{dz}{d\theta} = -1 + i \tan \tfrac{1}{2} \operatorname{Arg} z .$$

Obviously this defines a continuous ld system on the set P described above; the trajectories have equations

$$\varrho^3 = c \sin^{-2} \tfrac{2}{3}\varphi \quad (c > 0)$$

in polar coordinates ϱ, φ. It should be remarked that this example is a minor modification of the one used, for a different purpose, in Niemyckij and Stepanov (1949).

CHAPTER VI

Dynamical Systems on Special Carrier Spaces

The topological constitution of a carrier space exercises considerable influence on the permissible dynamical structures; a good instance is that compactness of the carrier implies globality of the dynamical system, IV,1.10. It may even be useful to classify continuous dynamical systems according to the topological properties of their carriers. Even though this is not the present intention, in this chapter there are presented some properties of continuous dynamical systems on the types of carrier space indicated in the section headings. Successive specialization is carried down to n-manifolds; the following two chapters then treat dynamical systems on 2-manifolds.

The variety of topics treated may make this chapter a little confusing to read. To alleviate this, the portions needed in the sequel will be indicated: Section 1 to item 1.8; Section 2 complete, but 2.14–21 are required only in VIII,4; in Section 3 only items 3.12–16; and items 4.6–8 in Section 4. The remaining subjects, in particular the third local determinacy theorem and geometric equivalence in Section 1, fundamentals of perturbation theory in Section 3, and properties of dynamical systems on manifolds with boundary in Section 4 are of general interest, since they indicate further intriguing problems.

As for references, the results up to 1.8 are possibly familiar, at least for differential gd systems or continuous gd systems on metrisable carriers: for example Niemyckij and Stepanov, 1949, § 1 for 1.2 and 1.6. The proof of 2.12 is based on Hájek 6, 1965.

1. DYNAMICAL SYSTEMS ON HAUSDORFF SPACES

The most immediate consequences of restricting the carriers of the continuous dynamical systems studied to T_2 spaces are the following. Limits of (generalized) convergent sequences are determined uniquely, and any subsequence of a convergent sequence has the same limit; quasi-compactness is, by definition, equivalent with compactness, and there cannot exist feebly critical points (cf. III,1.10). The condition for criticality of a point, II,4.6, can be weakened usefully; this is performed in the fundamental lemma 1.1, from which further consequences are then drawn.

1.1. Lemma. Let $\top$ *be a continuous lsd system on a* T_2 *space* P, *and* $x \in P$. *If, for some* (x_i, θ_i),

$$x_i \to x \quad P\,, \; in \; 0 < \theta_i \to 0 \; in \; \mathrm{R}^+\,, \quad x_i = x_i \top \theta_i\,,$$

then x *is a critical point.*

Proof. Take any $\theta \in [0, \alpha_x)$, and let n_i be the integral part of θ/θ_i. Then

$$0 \leqq n_i\theta_i \leqq \theta < n_i\theta_i + \theta_i\,,$$

so that $n_i\theta_i \to \theta$ in R^+. Then from the assumptions,

$$x \leftarrow x_i = x_i \top \theta_i = x_i \top n_i\theta_i \to x \top \theta\,;$$

and since limits are unique, $x = x \top \theta$ for all $\theta \in [0, \alpha_x)$. Hence from IV,4.6, x is a critical point as asserted.

If $x_i \equiv x$ is taken in 1.1, a new proof of non-existence of feebly critical points is obtained. Another consequence is:

1.2. Corollary. The set of critical points of a continuous lsd system on a T_2 *carrier is closed.*

1.3. Lemma. Let $\top$ *be a continuous lsd system on a* T_2 *space, and let* C_i *be cycles with periods* $\lambda_i \to 0$. *Then* limsup C_i *consists of critical points, and if* liminf $C_i \neq \emptyset$, lim C_i *is a single critical point.*

Proof. The first assertion is a special case of 1.1. For the second it suffices to show that $x \in$ liminf C_i and $y \in$ limsup C_i implies $x = y$.

Take $x_i \in C_i$ with $x_i \to x$, and $y_j \in C_j$ with $y_j \to y$ for some subsequence. Then $C_j = x_j \top [0, \lambda_j]$ by assumption, so that $y_j = x_j \top \theta_j$ with

$$0 \leqq \theta_j < \lambda_j \to 0\,.$$

Then

$$x \leftarrow x_j \top \theta_j = y_j \to y$$

and $x = y$ as was to be proved.

1.4. Corollary. Let $\top$ *be a continuous lsd system on a* T_2 *space* P, *and* $X \subset P$ *a compact subset without critical points. Then there is a positive lower bound to the primitive periods of cycles intersecting* X.

1.5. Lemma. Let $\top$ *be a continuous lsd system on a* T_2 *space* P. *Then, for each* $\lambda \in \mathrm{R}^+$, *the set* X_λ *consisting of all critical points and all points on cycles with periods* $\leqq \lambda$ *is closed in* P. *Hence if* P *is perfectly normal, the set of points on cycles is an* F_σ *in* P.

Proof. 1.5.1. Let $x_i \to x$ with each $x_i \in X_\lambda$, i.e. critical or on a cycle with period $\lambda_i \leqq \lambda$. If x_j is critical for some confinal indexes, then x is critical according to 1.2. In the opposite case there is either $\lambda_i \to 0$, whereupon x is again critical, from 1.3; or $\lambda_j \to \lambda'$ with $0 < \lambda' \leqq \lambda$ for some subsequence, whereupon

$$x \leftarrow x_j = x_j \mathsf{T} \lambda_j \to x \mathsf{T} \lambda',$$

so that x is either critical or on a cycle with period $\lambda' \leqq \lambda$. In all cases then $x \in X_\lambda$, proving the first assertion.

1.5.2. For the second, the set in question is, obviously,

$$\bigcup_{n=1}^{\infty} X_n - X_0 \tag{1}$$

with each X_n closed (1.5.1, 1.2) and X_0 the set of critical points. If P is perfectly normal, the open set $P - X_0$ is an F_σ, and thus the same holds of the set (1). This concludes the proof.

Since there are no feebly critical points, the classification II,4.6 of points in T_2 carriers reduces to the classical trichotomy:

1.6. Proposition. Let T *be a continuous lsd system with −unicity on a* T_2 *space* P, *and take any* $x \in P$. *Then there obtains precisely one of the following cases:*

1. x is a critical point.

2. x is on a cycle, and $x \mathsf{T} \mathsf{R}^+$ *is a simple closed curve.*

3. x is on a non-cyclic +trajectory, and the +solution t_x^+ *maps each compact subset of* $[0, \alpha_x)$ *homeomorphically into the non-compact set* $x \mathsf{T} \mathsf{R}^+ = \text{image } t_x^+$.
Similar results hold for continuous ld systems; in particular, each non-cyclic trajectory is non-compact.

Proof. III,1.10 may be applied to 4.6; in both 2−3 use t_x^+ to parametrize $x \mathsf{T} \mathsf{R}^+$, recalling that a continuous 1−1 map of a compact set into a T_2 space is a homeomorphism (Bourbaki, 1951, I, § 10, 4). Thus it only remains to prove the assertions on non-compactness in 3.

Assume that $x \mathsf{T} \mathsf{R}^+$ is compact; then $\alpha_x = +\infty$ from IV,1.10, so that the translation T_1 partializes to a homeomorphism

$$x \mathsf{T} \mathsf{R}^+ \approx x \mathsf{T} [1, +\infty).$$

Furthermore, from IV,4.7 it then follows that

$$\emptyset \neq L_x^+ \subset x \mathsf{T} [1, +\infty).$$

Since L_x^+ is invariant (cf. V,2.10), this implies $x \in L_x^+ \subset x \mathsf{T} [1, +\infty)$, so that $x = x \mathsf{T} \theta$ for some $\theta \geqq 1$, and thus 1 or 2 obtain. A similar proof applies to continuous ld systems.

1.7. It should be noted that in 1.6.3, t_x^+ can easily not be a homeomorphism $\mathsf{R}^+ \approx \approx x \mathsf{T} \mathsf{R}^+$, since one may have $x \mathsf{T} \theta_i \to x$ with $\theta_i \to +\infty$; a necessary and sufficient condition for $t_x^+ : \mathsf{R}^+ \approx x \mathsf{T} \mathsf{R}^+$ is that x is not Poisson +stable (cf. IV,4.8).

Next, some results will be obtained concerning sections and local determinacy.

1.8. Lemma. Let T *be a continuous lsd system on a* T_2 *space. If* Q_1, Q_2 *are disjoint sequentially compact sections, then* $Q_1 \cup Q_2$ *is also a section.*

Proof. 1.8.1. First show that there exists an $\varepsilon > 0$ with

$$(Q_1 \mathsf{T} \theta) \cap (Q_2 \mathsf{T} \theta') = \emptyset \quad \text{for all} \quad \theta, \theta' \text{ in } [0, \varepsilon]. \tag{2}$$

Assuming the contrary, there exist

$$x_n \in Q_1, \quad x_n' \in Q_2 \quad \text{and} \quad \theta_n \to 0,\ \theta_n' \to 0 \text{ in } \mathsf{R}^+$$

with

$$x_n \mathsf{T} \theta_n = x_n' \mathsf{T} \theta_n'. \tag{3}$$

Since the Q_k are sequentially compact, one may even assume that $x_n \to x$ in Q_1, $x_n' \to x'$ in Q_2, whereupon from (3)

$$x = x \mathsf{T} 0 \leftarrow x_n \mathsf{T} \theta_n = x_n' \mathsf{T} \theta_n' \to x'.$$

Thus $x = x'$, contradicting the assumed disjointness of the Q_k; this proves (2).

1.8.2. Now take $\delta > 0$ smaller than ε and both the t-extents of the Q_k. Then for $Q = Q_1 \cup Q_2$ and $0 \leqq \theta < \theta' \leqq \delta$ one has

$$\begin{aligned}(Q \mathsf{T} \theta) \cap (Q \mathsf{T} \theta') = ((Q_1 \mathsf{T} \theta) \cap (Q_1 \mathsf{T} \theta')) \cup ((Q_2 \mathsf{T} \theta) \cap (Q_2 \mathsf{T} \theta')) \cup \\ \cup ((Q_1 \mathsf{T} \theta) \cap (Q_2 \mathsf{T} \theta')) \cup ((Q_1 \mathsf{T} \theta') \cap (Q_2 \mathsf{T} \theta)) = \emptyset\end{aligned}$$

from II,6.5 in the first line, and from 1.8.1. in the second. Hence $Q_1 \cup Q_2$ is indeed a section of t-extent δ, concluding the proof of 1.8.

A re-organization of the proofs of 1.1 and III,2.1 yields a further local determinacy theorem; in comparison with those of III,2 it is required that both the lsd systems concerned be continuous on a T_2 carrier, but all the other assumptions are weakened considerably; in particular there are no requirements on −unicity or s-points, and the local equivalence condition is unilateral.

1.9. Third local determinacy theorem. Let T *and* T' *be continuous lsd (or ld) systems on a* T_2 *space* P *with the following weak equivalence condition: There exists a sequence* $\theta_n \in \mathsf{R}^+$ *with* $0 < \theta_n \to 0$ *such that, for all* $x \in P$,

$$x \mathsf{T} \theta_n = x \mathsf{T}' \theta_n \tag{4}$$

whenever both terms are defined. Then $\mathsf{T} = \mathsf{T}'$.

Proof. 1.9.1. From (4) it follows immediately that

$$x \mathsf{T} k\theta_n = x \mathsf{T}' k\theta_n \tag{5}$$

for all $k \in \mathsf{C}^+$, provided both terms are defined.

1.9.2. Now show that

$$x \mathsf{T} \theta = x \mathsf{T}' \theta \quad \text{for} \quad x \in P,\ 0 \leqq \theta < \min(\alpha_x, \alpha'_x). \tag{6}$$

To any such x, θ, let k_n be the integral part of θ/θ_n; then

$$0 \leqq k_n\theta_n \leqq \theta \leqq k_n\theta_n + \theta_n, \quad k_n\theta_n \to \theta \text{ in } \mathsf{R}^+,$$

and in particular, both $x \mathsf{T} k_n\theta_n$ and $x \mathsf{T}' k_n\theta_n$ are defined for large n. Hence from (5),

$$x \mathsf{T} \theta \leftarrow x \mathsf{T} k_n\theta_n = x \mathsf{T}' k_n\theta_n \to x \mathsf{T}' \theta,$$

proving (6).

1.9.3. Repeat the proof of III,2.1.6, using the given T_2 topology of P to replace the inherent topology and applying IV,1.4. This yields $\alpha_x = \alpha'_x$ for all $x \in P$, and completes the proof of 1.9 in the semi-dynamical case.

1.9.4. Now assume that both T, T' are continuous ld systems on P. Passing to their semi-dynamical restrictions, it has already been proved that $\alpha_x = \alpha'_x$ always and

$$x \mathsf{T} \theta = x \mathsf{T}' \theta \quad \text{for} \quad 0 \leqq \theta < \alpha_x.$$

Keeping $x \in P$ fixed, take any $\theta \in \mathsf{R}^1$ with

$$\max(\beta_x, \beta'_x, -\alpha_x) < \theta \leqq 0; \tag{7}$$

then $x \mathsf{T} \theta = y$ is defined, and

$$x = y \mathsf{T}(-\theta) = y \mathsf{T}'(-\theta),$$

proving that

$$x \mathsf{T} \theta = y = x \mathsf{T}' \theta.$$

With (7), this is a condition stronger than (4) for the lsd restrictions of the orientation-changed T, T'; application of 1.9.1–3 then completes the proof of 1.9.

1.10. Corollary. For continuous lsd systems on a T_2 space P, $\mathsf{T} \subset \mathsf{T}'$ implies $\mathsf{T} = \mathsf{T}'$; hence a non-global system cannot be augmented to a global system on P.

1.11. Remarks. From 1.9 it follows that a continuous lsd system on a T_2 carrier P is completely determined by its behaviour on a rather meager set in $P \times \mathsf{R}^+$. A case of particular interest arises on taking for T' in 1.9 a differential ld system. Thus, if T is a continuous ld system on an open set $G \subset \mathsf{R}^m$, and if for some sequence $0 < \theta_n \to 0$ in R^+ at least and all $x \in G$,

$$\lim_n \frac{1}{\theta_n} (x \mathsf{T} \theta_n - x) = f(x)$$

exists and defines a continuous map $f : G \to \mathsf{R}^m$, and finally, if the differential equation in G

$$\frac{dx}{d\theta} = f(x) \tag{8}$$

has unicity of solutions, T is precisely the differential ld system defined by (8). A sufficient condition for both existence and continuity of f and unicity in (8) is

$$\|(x \mathsf{T} \theta_n - y \mathsf{T} \theta_n) - (x - y)\| \leqq \lambda \|x - y\| \, \theta_n$$

for some $\lambda \in \mathsf{R}^+$ and all x, θ_n. In general, it may happen that all these conditions are satisfied except for unicity.

1.12. Example. Let T be defined by

$$x \mathsf{T} \theta = (\sqrt[3]{x} + \tfrac{1}{3}\theta)^3 \quad \text{for} \quad x \in \mathsf{R}^1,\ \theta \in \mathsf{R}^1 . \tag{9}$$

It is easily verified that T is a continuous gd system on R^1; and

$$\lim_{\theta \to 0+} \frac{1}{\theta} (x \mathsf{T} \theta - x) = \lim_{\theta \to 0} \frac{1}{\theta} ((\sqrt[3]{x})^2 \, \theta + \tfrac{1}{3} \sqrt[3]{(x)} \, \theta^2 + \tfrac{1}{27}\theta^3) = (x^2)^{1/3}$$

does define a continuous map $\mathsf{R}^1 \to \mathsf{R}^1$. However, the corresponding differential equation does not have unicity of solutions, see I,1.6.

It may be noted that T is isomorphic in $\mathcal{D}_g^T$ to the differential gd system defined by $dx/d\theta = \frac{1}{3}$ via the homeomorphism $f : \mathsf{R}^1 \approx \mathsf{R}^1$, $f(x) = x^3$; this follows directly from (9). Since the latter differential equation does have unicity of solutions, there is no concept in dynamical system theory corresponding to that of a singular solution.

1.13. In differential equation theory, two differential ld systems on the same carrier $G \subset \mathsf{R}^n$,

$$\frac{dx}{d\theta} = f(x) , \quad \frac{dx}{d\theta} = g(x) , \tag{10}$$

(with assumptions as in IV, 1.3.2) are called *geometrically equivalent* if each solution of either system is the re-parametrization of some solution of the other; this formulates a definition implicit in Lefschetz (1957). It is fairly easily shown that this condition obtains iff there exists a real-valued map φ with

$$f(x) = \varphi(x) \, g(x) , \quad \varphi(x) \neq 0$$

whenever $x \in G$ and $g(x) \neq 0$ (then obviously both φ and $1/\varphi$ are continuous; φ is closely related to the derivative of the re-parametrizing function). The abstract analogue of this concept applying to continuous ld systems will now be presented.

1.14. Theorem. If T *and* T' *are continuous ld systems on a* T_2 *space* P*, then the following properties are equivalent:*

1. To each $x \in P$ *there is a homeomorphism* $\varphi_x : (\beta_x, \alpha_x) \approx (\beta'_x, \alpha'_x)$ *such that*

$$x \,\mathsf{T}\, \theta = x \,\mathsf{T}'\, \varphi_x(\theta) \tag{11}$$

whenever either side is defined.

2. The inherent topologies defined by T *and* T' *on* P *coincide.*

3. Each semi-trajectory of T *coincides with some semitrajectory of* T'*, and conversely.*

If in 1 it is also required that $\varphi_x(0) = 0$*, then, for non-critical* $x \in P$*,* φ_x *is determined uniquely.*

Proof. 1.14.1. That 1 implies 2 is verified easily: for each $x \in P$ and sufficiently small $\varepsilon > 0$,

$$x \,\mathsf{T}\,(-\varepsilon, \varepsilon) = x \,\mathsf{T}'\,(\varphi_x(\mp\varepsilon), \varphi_x(\pm\varepsilon))$$

is a basic open neighbourhood for both the inherent topologies; cf. III,1.2.

1.14.2. Now prove that 2 implies 3. In terms of the inherent topology, the trajectories are precisely the t-components of P (cf. III,1.2); hence if the inherent topologies of T and T' coincide, so do their trajectories. In particular, the two ld systems have coinciding critical points, cycles, and non-cyclic trajectories, since these are precisely the singelton, the t-compact and the non-t-compact t-components; and of course, according to 1.6, there are no further types of trajectories.

For critical points and cycles, semi-trajectories are simply trajectories; thus to prove 3 it only remains to show that the non-cyclic semi-trajectories of T and T' coincide. In this case a trajectory $x \,\mathsf{T}\, \mathsf{R}^1$, in its inherent topology, is homeomorphic to R^1, so that

$$x \,\mathsf{T}\, \mathsf{R}^1 - (x) = (x \,\mathsf{T}\,(-\infty, 0)) \cup (x \,\mathsf{T}\,(0, +\infty))$$

is the decomposition into components. For like reason

$$x \,\mathsf{T}\, \mathsf{R}^1 - (x) = x \,\mathsf{T}'\, \mathsf{R}^1 - (x) = (x \,\mathsf{T}'\,(-\infty, 0)) \cup (x \,\mathsf{T}'\,(0, +\infty)) \tag{12}$$

is again the decomposition into components. Hence $x \,\mathsf{T}\,(0, +\infty)$ coincides with one of the summands of (12), for example with $x \,\mathsf{T}'\,(0, +\infty)$, and then

$$x \,\mathsf{T}\, \mathsf{R}^+ = x \,\mathsf{T}'\, \mathsf{R}^+\,, \quad x \,\mathsf{T}\, \mathsf{R}^- = x' \,\mathsf{T}\, \mathsf{R}^-\,. \tag{13}$$

Observe that this argument also yields the following. If (13) holds with $x \mathsf{T} \mathsf{R}^+$ non-cyclic, then

$$(x \mathsf{T} \theta) \mathsf{T} \mathsf{R}^+ = (x \mathsf{T} \theta) \mathsf{T}' \mathsf{R}^+ \tag{14}$$

if $x \mathsf{T} \theta$ is defined.

1.14.3. The proof of 3 ⇒ 1 is slightly more involved. Again, the two ld systems have coinciding critical points, cycles, and non-cyclic semi-trajectories, since according to 1.6 these are singletons, compact and non-compact respectively. Take any $x \in P$; it may be assumed, if necessary by changing orientation, that $x \mathsf{T} \mathsf{R}^+ = x \mathsf{T}' \mathsf{R}^+$. If x is critical, one may choose $\varphi_x(\theta) \equiv \theta$ to satisfy (11).

1.14.4. Next, assume that $x \mathsf{T} \mathsf{R}^+$ is a cycle, with period say τ in T. Then the solution t_x may be factorized, $t_x = s_x p$, with p the exponential map $p(\theta) = \exp 2\pi i\theta/\tau$ and s_x a homeomorphism:

$$t_x : \mathsf{R}^1 \xrightarrow{p} \mathsf{R}^1 \bmod \tau \overset{s}{\approx} x \mathsf{T} \mathsf{R}^1 = x \mathsf{T} \mathsf{R}^+ .$$

Similarly, of course, for the second ld system T'; there results the diagram

$$\begin{array}{ccccc} t_x : \mathsf{R}^1 & \xrightarrow{p} & \mathsf{R}^1 \bmod \tau & \overset{s_x}{\approx} & x \mathsf{T} \mathsf{R}^+ \\ \downarrow \varphi_x & & \downarrow \psi_x & & \| \\ t'_x : \mathsf{R}^1 & \xrightarrow[p']{} & \mathsf{R}^1 \bmod \tau' & \underset{s_x'}{\approx} & x \mathsf{T}' \mathsf{R}^+ \end{array}$$

where ψ_x is the homeomorphism defined by requiring commutativity in the latter square, and φ_x the covering homeomorphism of ψ_x induced by the covering maps p and p'. Hence

$$t_x = s_x p = s'_x p' \varphi_x = t'_x \varphi_x$$

or

$$x \mathsf{T} \theta = t_x(\theta) = t'_x(\varphi_x(\theta)) = x \mathsf{T}' \varphi_x(\theta) \tag{15}$$

as required in (11).

1.14.5. Finally let $x \mathsf{T} \mathsf{R}^+ = x \mathsf{T}' \mathsf{R}^+$ be non-cyclic. This case is slightly more complicated since the solutions, continuous and $1-1$ onto, are not necessarily homeomorphic. One has the diagram

$$\begin{array}{ccc} t_x : (\beta_x, \alpha_x) & \to & x \mathsf{T} \mathsf{R}^1 \\ \downarrow \varphi_x & & \| \\ t'_x : (\beta'_x, \alpha'_x) & \to & x \mathsf{T}' \mathsf{R}^1 . \end{array}$$

Here φ_x may again be defined by requiring commutativity, since the solutions are $1-1$ onto. In particular then φ_x is $1-1$ onto, and (11) holds as in (15); it remains to prove that φ_x is homeomorphic. Take any θ_1, θ_2 with $\beta_x < \theta_1 < \theta_2 < \alpha_x$; then from $x \mathsf{T} \mathsf{R}^+ = x \mathsf{T}' \mathsf{R}^+$ and (14) it follows that

$$x \mathsf{T} [\theta_k, +\infty) = x \mathsf{T}' [\varphi_x(\theta_k), +\infty),$$

and then

$$\begin{array}{ccc} x \mathsf{T} [\theta_1, +\infty) & \supset & x \mathsf{T} [\theta_2, +\infty) \\ \| & & \| \\ x \mathsf{T}' [\varphi_x(\theta_1), +\infty) & \supset & x \mathsf{T}' [\varphi_x(\theta_2), +\infty). \end{array}$$

Hence $\varphi_x(\theta_1) < \varphi_x(\theta_2)$, proving that φ_x is strictly monotone. Since it maps onto the interval (β'_x, α'_x), φ_x is continuous, and hence a homeomorphism. This completes the proof that 3 implies 1.

1.14.6. It follows directly from $-$unicity and continuity of φ_x that, for non-critical $x \in P$, φ_x is uniquely determined by (11) and $\varphi_x(0) = 0$. This concludes the proof of 1.14.

1.15. Remarks. In 1.14.1, from the group property and (11), it follows that

$$\varphi_x(\theta + \theta') = \varphi_x(\theta) + \varphi_{x \mathsf{T} \theta}(\theta')$$

for non-critical $x \in P$ (whenever $\varphi_x(\theta)$ and either side is defined). In particular, if $\varphi_{x \mathsf{T} \theta}$ does not depend on θ, then φ_x is linear continuous, and therefore in (11)

$$x \mathsf{T} \theta = x \mathsf{T}' (\psi(x)\, \theta)$$

with $0 \neq \psi(x) \in \mathsf{R}^1$. A simple example is orientation change, $\psi(x) \equiv -1$; also see II,3.11.

Geometric equivalence, as defined by the properties in 1.14, is an equivalence relation on $\mathscr{D}_l^T(P)$ finer than isomorphism. As examples of "geometric" concepts, one may list critical points, cycles, trajectories, invariance, t-convexity, orbital stability; however, semi-trajectories, periods of cycles, $+$invariance, $o+$ stability are not geometric. Another illustration of geometric equivalence is the following theorem due to Vinograd (Niemyckij and Stepanov, I, § 3): Each differential gd system with carrier G connected and open in R^n is geometrically equivalent to the relativization of a differential gd system in R^n itself.

2. DYNAMICAL SYSTEMS ON TYCHONOV SPACES

2.1. An alternative heading might well have been "Extension of Sections". Since each subset of a section is a section on its own account (II,6.9.1), it is natural to inquire about the possibility of extending a given section S. However, even the closure of S need not be a section (II,6.6.3, with the natural topology for R^2); and according to 1.8, the union of disjoint compact sections is always a section. Therefore attention centres more around the closed and connected sections. The requirement that the extension $S' \supset S$ be as large as possible will be interpreted in the sense that some $S' \mathsf{T} (-\delta, \delta)$ is to be a neighbourhood of S; in 4.7 it is then shown that this condition excludes situations such as both S and S' being simple arcs in R^3.

2.2. A construction will now be performed, in separately numbered steps. In the course of this, up to 2.12, there is assumed given a continuous ld system T on a T_ϱ space P, and a compact section S of T. Two significant objects, denoted as U and G, will be considered subject to certain conditions, but otherwise left free to vary; a subsequent choice of these will yield further properties of the extended section $S' \supset S$ to be constructed.

Merely for the purposes of this section, the following convention will be useful: if $J = [\alpha_1, \alpha_2]$ is an interval in R^1, then let

$$\lambda J = [\lambda\alpha_1, \lambda\alpha_2] \quad \text{for} \quad \lambda > 0\,, \quad J^\circ = (\alpha_1, \alpha_2)\,.$$

2.3. Take any $\varepsilon > 0$ such that

$$4\varepsilon \leqq \textit{t-extent of } S\,, \quad S \times 2J \subset \textit{domain}\ \mathsf{T} \tag{1}$$

for $J = [-\varepsilon, \varepsilon]$; according to IV,2.1, such ε's indeed exist. Let

$$S \mathsf{T}\, 2J \xrightarrow{h} S \times 2J \xrightarrow{p} 2J\,,$$

where h is the homeomorphism inverse to $\mathsf{T} \mid (S \times 2J)$ (cf. IV, 2.17) and p the "second" projection map. Then

$$ph : S \mathsf{T}\, 2J \to 2J \subset \mathsf{R}^1$$

is a continuous real-valued map whose domain $S \mathsf{T}\, 2J$ is compact in P, according to IV,2.8. Hence there exists a continuous extension ψ of ph,

$$ph \subset \psi : P \to \mathsf{R}^1\,.$$

(This is simple but possibly not quite standard: compactify the T_ϱ space P, obtaining a normal space βP with $S \mathsf{T}\, 2J$ as compact subset; now extend ph via the Tietze-Urysohn theorem (Bourbaki, 1948, § 1, 5 and § 4, 2), and then restrict it back to $P \subset \beta P$.) To summarize: There exists a continuous map $\psi : P \to \mathsf{R}^1$ such that, for $(x, \theta) \in S \times 2J$,

$$\psi(x \mathsf{T} \theta) = ph(x \mathsf{T} \theta) = p(x, \theta) = \theta\,. \tag{2}$$

Obviously $\psi(x \mathsf{T} \theta)$ is a continuous function of x and θ; this – and similar remarks below – should be interpreted thus: The composed map $\psi\, \mathsf{T}$,

$$\psi\, \mathsf{T}\, (x, \theta) = \psi(x \mathsf{T} \theta)\,,$$

is continuous.

2.4. Since S is compact, each neighbourhood contains a closed neighbourhood (this will be applied again in 2.5); hence and from (1) and IV,2.2, there exists a neighbourhood U of S such that $\bar{U} \times 2J \subset \textit{domain}\ \mathsf{T}$. A continuous map $\varphi : \bar{U}\ \mathsf{T}\ J \to \mathsf{R}^1$ may then be defined by

$$\varphi(x) = \frac{1}{2\varepsilon}\int_{-\varepsilon}^{\varepsilon} \psi(x\ \mathsf{T}\ \lambda)\, \mathrm{d}\lambda\,.$$

In turn there follow these properties of φ:

$$\varphi(x\ \mathsf{T}\ \theta) = \frac{1}{2\varepsilon}\int_{-\varepsilon}^{\varepsilon} \psi(x\ \mathsf{T}\ (\theta + \lambda))\, \mathrm{d}\lambda = \frac{1}{2\varepsilon}\int_{\theta-\varepsilon}^{\theta+\varepsilon} \psi(x\ \mathsf{T}\ \lambda)\, \mathrm{d}\lambda\,,$$

$$\frac{\partial}{\partial\theta}\varphi(x\ \mathsf{T}\ \theta) = \frac{1}{2\varepsilon}\big(\psi(x\ \mathsf{T}\ (\theta + \varepsilon)) - \psi(x\ \mathsf{T}\ (\theta - \varepsilon))\big)$$

for $(x, \theta) \in \bar{U} \times J$, both continuous functions of (x, θ). In particular, for $x \in S$ one may apply (2) to obtain

$$\varphi(x\ \mathsf{T}\ \theta) = \theta\,, \quad \frac{\partial}{\partial\theta}\varphi(x\ \mathsf{T}\ \theta) = 1\,, \tag{3}$$

$$\varphi(x) = 0\,, \tag{4}$$

for all $(x, \theta) \in S \times J$.

2.5. Next, some neighbourhoods of S will be chosen. From (3) and continuity, there exists a neighbourhood V_1 of S with

$$V_1 \subset U\,, \quad \frac{\partial}{\partial\theta}\varphi(x\ \mathsf{T}\ \theta) > 0 \quad \text{for} \quad (x\ \ \theta) \in V_1 \times J\,. \tag{5}$$

From compactness of S, there exists a neighbourhood V_2 of S and a $\delta > 0$ such that for $I = [-\delta\ \ \delta]$,

$$V_2\ \mathsf{T}\ 2I \subset V_1\,, \quad \delta \leqq \varepsilon\,;$$

in particular, $I \subset J$. Using this value of δ, from (3) there follows

$$\varphi(x\ \mathsf{T}\ \delta) = \delta > 0 > -\delta = \varphi(x\ \mathsf{T}\ (-\delta)) \quad \text{for} \quad x \in S\,.$$

Thus, finally, there exists an open neighbourhood G of S such that $\bar{G} \subset V_2$ and

$$\varphi(x\ \mathsf{T}\ \delta) > 0 > \varphi(x\ \mathsf{T}\ (-\delta)) \quad \text{for} \quad x \in \bar{G}\,. \tag{6}$$

To summarize these results for U, G and $I = [-\delta, \delta]$: the set G is open and

$$S \subset G \subset \bar{G}\ \mathsf{T}\ 2I \subset U\,, \quad \bar{U} \times 2I \subset \textit{domain}\ \mathsf{T}\,, \tag{7}$$

$$\frac{\partial}{\partial\theta}\varphi(x\ \mathsf{T}\ \theta) > 0 \quad \text{for} \quad (x, \theta) \in \bar{G} \times 3I \tag{8}$$

(since then $x \mathrm{T} \theta \in (\bar{G} \mathrm{T} 2I) \mathrm{T} I \subset V_1 \mathrm{T} J$ and (5) applies). This concludes the preliminaries.

2.6. Lemma. The set

$$S' = \{x : \varphi(x) = 0\} \cap (\bar{G} \mathrm{T} I) \tag{9}$$

is a closed section of t-extent 2δ and with $S' \times I \subset$ domain T.

Proof. 2.6.1. To show that $\bar{G} \mathrm{T} I$ is closed, let $(x_i\ \theta_i) \in \bar{G} \mathrm{T} I$ with $x_i \mathrm{T} \theta_i \to x'$. With I compact, it may be assumed that $\theta_i \to \theta \in I$; and since

$$x' \in \overline{\bar{G} \mathrm{T} I} \subset \bar{U}\,,$$

it follows from the second relation in (7) that $x' = x \mathrm{T} \theta$ for some $x \in P$. Then

$$x_i = (x_i \mathrm{T} \theta_i) \mathrm{T} (-\theta_i) \to (x \mathrm{T} \theta) \mathrm{T} (-\theta) = x \in \bar{G}\,,$$

proving that

$$\lim x_i \mathrm{T} \theta_i = x' = x \mathrm{T} \theta \in \bar{G} \mathrm{T} I$$

as required.

2.6.2. Since φ is continuous, its zero point set is closed in domain φ; from (7), $\bar{G} \mathrm{T} I \subset U \subset$ *domain* φ with $\bar{G} \mathrm{T} I$ closed, as has just been shown. Thus S' is a closed subset of P.

2.6.3. Now show that S' is a section of t-extent 2δ. To this end take any $x \in S'$ and $\theta \in 2I$ with $0 \neq \theta$ and aim at proving $x \mathrm{T} \theta \notin S'$ (cf. II,6.8.2). From (9) there follows $x \in \bar{G} \mathrm{T} I$, so that $x = x' \mathrm{T} \theta'$ for suitable $(x', \theta') \in \bar{G} \times I$, and then

$$x \mathrm{T} \theta = x' \mathrm{T} (\theta' + \theta)\,, \quad \theta' + \theta \in 3I\,.$$

It follows that

$$\varphi(x \mathrm{T} \theta) - \varphi(x) = \varphi(x' \mathrm{T} (\theta' + \theta)) - \varphi(x' \mathrm{T} \theta') = \int_{\theta'}^{\theta' + \theta} \frac{\partial}{\partial \lambda} \varphi(x' \mathrm{T} \lambda)\, \mathrm{d}\lambda$$

and according to (8) the last term is non-zero, having the same sign as $\theta \neq 0$. Since $x \in S'$ by assumption, one has $\varphi(x) = 0$, and hence $\varphi(x \mathrm{T} \theta) \neq 0$ and $x \mathrm{T} \theta \notin S'$ as required.

2.6.4. Finally, $S' \times I \subset$ *domain* T follows from $S' \subset \bar{G} \mathrm{T} I$ and (7). This completes the proof of 2.6.

2.7. Lemma. $S \subset G \subset \bar{G} \subset S' \mathrm{T} I^\circ$.

Proof. $S \subset G$ is among (7). For the last inclusion, take any $x \in G$; from (6) and continuity of φ, one has $\varphi(x \mathrm{T} \theta) = 0$ for suitable $\theta \in I^\circ$. Obviously then $x \mathrm{T} \theta \in \bar{G} \mathrm{T} I$, so that $x \mathrm{T} \theta \in S'$ by construction (9), and conversely

$$x = (x \mathrm{T} \theta) \mathrm{T} (-\theta) \in S' \mathrm{T} I^\circ\,.$$

2.8. Lemma. $S \subset S' \subset \bar{G} \top I \subset U$.

Proof. The first inclusion follows from (4) and (9), using $S \subset G$ from (7); the second inclusion follows from (9) directly; and the last from (7) again.

2.9. Since G is open, 2.7 implies that $S' \top (-\delta, \delta)$ is a neighbourhood of S; and $S' \subset U$ in 2.8 shows that S' may be made to lie arbitrarily near to S (cf. the construction of U in 2.4).

Having established 2.6, the results of IV, 2.17–18 may be applied to S' and 2δ. Therefore one has available a retraction mapping $r : S' \times I \to S'$ with

$$r(x \top \theta) = x \quad \text{for} \quad (x, \theta) \in S' \times I \tag{10}$$

2.10. Lemma. r is a closed map taking $\bar{G} \subset S' \top I$ onto S'.

Proof. 2.10.1. First it is required to show that $r(F)$ is closed whenever F is closed in $S' \top I$. Thus let $r(F) \ni x_i \to x$, and attempt to prove that also $x \in r(F)$. One has $x \in S'$, and also $x_i \top \theta_i \in F$ for suitable $\theta_i \in I$. With I compact, it may be assumed that $\theta_i \to \theta \in I$, and then

$$x_i \top \theta_i \to x \top \theta \in S' \top I .$$

As F is closed in $S' \top I$, there follows $x \top \theta \in F$, and indeed from (10)

$$x = r(x \top \theta) \in r(F) .$$

2.10.2. To show that r maps $\bar{G}$ onto S', take any $x \in S'$; from $S' \subset \bar{G} \top I$ in 2.8 there follows $x = x' \top \theta'$ for suitable $(x' \ \theta') \in \bar{G} \times I$. Hence $x' = x \top (-\theta') \in S' \top I$, and from (10)

$$x = r(x \top (-\theta')) = r(x') \in r(\bar{G}) .$$

This proves $S' \subset r(\bar{G})$; since obviously $r(\bar{G}) \subset r(S \top I) = S'$ from 2.7, this concludes the proof of 2.10.

2.11. Some further results from topology will be needed now. To recall a definition, a metric space M is said to have *property* $\mathcal{S}$ if, for every $\varepsilon > 0$, M is the union of a finite system of connected sets with diameters at most ε (Whyburn, 1942). Then

2.11.1. In a metric space with property $\mathcal{S}$ one may choose an open-set basis consisting of open sets with locally connected closures (*loc. cit*, I. 15.41 and 15.1).

The following assertion can be proved readily.

2.11.2. The image of a locally connected set under a continuous and closed map is again locally connected (Hall and Spencer, 1955, V, 5.2).

2.12. Theorem. Let $\top$ be a continuous ld system on a T_ϱ space P, and S any compact section. Then there exists a closed section S', of t-extent $2\delta > 0$, such that *$S \subset S'$ and that $S' \top (-\delta, \delta)$ is an arbitrarily small neighbourhood of S. Further-*

more, if P is locally compact, S′ may be taken compact; if P is locally connected and S connected then S′ may be taken connected; if P is locally metrisable with property $\mathscr{S}$, then S′ may be taken locally connected.

Proof. The first part of the assertion was proved in 2.6–8. Now consider any open-set basis $\mathscr{G}$ in P. Since S is compact, in 2.5 one may choose G of the form

$$G = \bigcup_{k=1}^{n} G_k\,, \quad G_k \in \mathscr{G}\,,$$

whereupon, using the retraction r from 2.9-10

$$\bar{G} = \bigcup_{1}^{n} \bar{G}_k \quad \text{and} \quad S' = r(\bar{G})\,. \tag{11}$$

If P is locally compact, then one may take for $\mathscr{G}$ the system of all open subsets with compact closures; and then in (11), $\bar{G}_k$, $\bar{G}$ and $S' = r(\bar{G})$ are all compact. If P is locally connected and S connected, then one may take all G_k connected, where upon

$$\bar{G} = \bigcup_{1}^{n} \bar{G}_k = S \cup \bigcup_{1}^{n} \bar{G}_k$$

is also connected and the same then holds of $S' = r(\bar{G})$. If P is locally metrisable with property $\mathscr{S}$, one may take $\bar{G}_k$ locally connected (cf. 2.11.1); then $\bar{G}$ is locally connected, so that $S' = r(\bar{G})$ is also such according to 2.11.2 and 2.10. This concludes the proof of 2.12.

2.13. Remarks. From the preceding proof it is evident that if P satisfies any combination of the three conditions, then S' may be taken with the corresponding combination of properties. Also note that, in the very last assertion, local connectedness is not required of S, so that S' inherits this property from the carrier space P and not from S; this will be exploited in VII,1.

In the principal application intended at present, the carrier P is an n-manifold with boundary; thus each point of P has a neighbourhood homeomorphic to E^n, and all the conditions on P mentioned in 2.12 are satisfied *a fortiori*.

Returning to the general situation of a T_ϱ space as carrier, every non-critical point is a compact section, according to II,6.9.5 and III,1.10.1–2. On applying 2.12 there results the following generalization of the Whitney-Bebutov theorem (Niemyckij and Stepanov, 1949, V, § 2), originally formulated for continuous gd systems on metrisable carriers. If x is a non-critical point of a continuous ld system on a T_ϱ space, then there exists a closed section $S' \ni x$ with $S' \mathrm{T} (-\delta, \delta)$ an arbitrarily small neighbourhood of x.

2.14. In Chapter VIII there will be needed another existence theorem for sections, which will now be established by a modification of the technique used in 2.3–12.

The situation may be described in general terms as follows. Within the carrier of a dynamical system T there is given a set Q which satisfies condition (12) below for some $\varepsilon > 0$. Obviously, if Q itself were a section, then (12) would be satisfied for all ε sufficiently small; however, in the intended application ε will be taken large. It is required to construct a section within $Q \mathsf{T} \mathsf{R}^+$ which is to separate Q from $Q \mathsf{T} \varepsilon$. Precise assumptions and the construction will now be presented.

2.15. Up to 2.21 there is assumed given a continuous gsd system T with $-$unicity on a perfectly normal space P; and also a compact subset $Q \subset P$ and an $\varepsilon > 0$ such that

$$Q \cap (Q \mathsf{T} [\varepsilon, 3\varepsilon]) = \emptyset . \tag{12}$$

For convenience, set

$$J = [\varepsilon, 3\varepsilon], \quad I = [0, \varepsilon];$$

as in 2.2, $2I$ will denote $[0, 2\varepsilon]$, etc.

2.16. From the assumptions and IV,2.8 it follows that Q and $Q \mathsf{T} J$ are disjoint compact; hence there exists a "separating" function, i.e. a continuous map $\psi_1 : P \to \mathsf{R}^1$ with $0 \leqq \psi_1 \leqq 1$ and

$$\psi_1 \mid Q = 0, \quad \psi_1 \mid (Q \mathsf{T} J) = 1 .$$

Since P is perfectly normal and $Q \mathsf{T} J$ closed, there exists a second continuous map $\psi_2 : P \to \mathsf{R}^1$ with $0 \leqq \psi_2 \leqq 1$ again, and

$$\psi_2(x) = 0 \quad \text{iff} \quad x \in Q \mathsf{T} J .$$

(For example, if P is metrisable, and d an admissible metric, then one could take $\psi_2(x) = \max(1, d(x, Q \mathsf{T} J))$.) Now set $\psi = \psi_1 . (1 - \psi_2)$, obtaining a continuous map $\psi : P \to \mathsf{R}^1$ with the following properties:

$$0 \leqq \psi \leqq 1, \quad \psi \mid Q = 0,$$
$$\psi(x) = 1 \quad \text{iff} \quad x \in Q \mathsf{T} J . \tag{13}$$

2.17. The next step is analogous to 2.4: define a continuous map $\varphi : P \to \mathsf{R}^1$ by

$$\varphi(x) = \frac{1}{\varepsilon} \int_0^\varepsilon \psi(x \mathsf{T} \lambda) \, d\lambda .$$

From (13) and continuity of ψ there follows easily

$$0 \leqq \varphi \leqq 1, \quad \varphi \mid (Q \mathsf{T} [\varepsilon, 2\varepsilon]) = 1, \tag{14}$$
$$\varphi(x) = 1 \quad \text{implies} \quad x \mathsf{T} I \subset Q \mathsf{T} J . \tag{15}$$

In particular, from (12), $\varphi(x) < 1$ for all $x \in Q$. Since Q is compact, this yields $\max_Q \varphi < 1$. Now choose any $\mu \in \mathsf{R}^1$ with

$$\max_Q \varphi < \mu < 1 \tag{16}$$

and set

$$S = \{x : \varphi(x) = \mu\} \cap (Q \mathbin{\mathsf{T}} 2I). \tag{17}$$

2.18. *Lemma. For fixed $x \in Q$ and θ varying in $2I$, $\varphi(x \mathbin{\mathsf{T}} \theta)$ is a non-decreasing function of θ, strictly increasing at each $\theta \in 2I$ with $\varphi(x \mathbin{\mathsf{T}} \theta) < 1$. To each $x \in Q$ there is a unique $\theta \in 2I$ with $x \mathbin{\mathsf{T}} \theta \in S$, whereupon $\theta \in I^\circ$.*

Proof. From the construction of φ there follows in turn,

$$\varphi(x \mathbin{\mathsf{T}} \theta) = \frac{1}{\varepsilon}\int_0^\varepsilon \psi(x \mathbin{\mathsf{T}} (\theta + \lambda))\, d\lambda = \frac{1}{\varepsilon}\int_\theta^{\theta+\varepsilon} \psi(x \mathbin{\mathsf{T}} \lambda)\, d\lambda,$$

$$\frac{\partial}{\partial\theta}\varphi(x \mathbin{\mathsf{T}} \theta) = \frac{1}{\varepsilon}\left(\psi(x \mathbin{\mathsf{T}} (\theta + \varepsilon)) - \psi(x \mathbin{\mathsf{T}} \theta)\right).$$

Now, for $(x, \theta) \in Q \times 2I$ one has $\theta + \varepsilon \in J$, whereupon from (13)

$$\psi(x \mathbin{\mathsf{T}} (\theta + \varepsilon)) = 1 \geqq \psi(x \mathbin{\mathsf{T}} \theta);$$

thus $\varphi(x \mathbin{\mathsf{T}} \theta)$ has a non-negative derivative, and indeed non-decreases. Furthermore, $\varphi(x \mathbin{\mathsf{T}} \theta) < 1$ then implies $x \mathbin{\mathsf{T}} \theta \notin Q \mathbin{\mathsf{T}} J$ according to (15), so that $\psi(x \mathbin{\mathsf{T}} \theta) < 1$ from (13); thus $\varphi(x \mathbin{\mathsf{T}} \theta)$ has a positive derivative.

The second assertion follows directly from (16), (14), continuity of φ and the result just proved. This establishes 2.18.

2.19. *Lemma. S is a compact section with t-extent ε and*

$$S \subset Q \mathbin{\mathsf{T}} I^\circ, \quad Q \mathbin{\mathsf{T}} \varepsilon \subset S \mathbin{\mathsf{T}} I^\circ. \tag{18}$$

Furthermore, S separates Q from $Q \mathbin{\mathsf{T}} [\varepsilon, 2\varepsilon]$ in $Q \mathbin{\mathsf{T}} 2I$.

Proof. Compactness of S follows from that of $Q \mathbin{\mathsf{T}} 2I$ and continuity of φ in (17). Using the formulation of II,6.7.2, S is a section of t-extent ε according to 2.18; this also yields the first relation in (18). For the second, take any $x \in Q \mathbin{\mathsf{T}} \varepsilon$, so that $x = x' \mathbin{\mathsf{T}} \varepsilon$ with $x' \in Q$. Then $x' \mathbin{\mathsf{T}} \theta \in S$ for a suitable $\theta \in I^\circ$, and

$$x = x' \mathbin{\mathsf{T}} \varepsilon = (x' \mathbin{\mathsf{T}} \theta) \mathbin{\mathsf{T}} (\varepsilon - \theta) \in S \mathbin{\mathsf{T}} I^\circ$$

as required.

Finally, (17) implies that

$$(Q \mathbin{\mathsf{T}} 2I) - S = \{x \in Q \mathbin{\mathsf{T}} 2I : \varphi(x) < \mu\} \cup \{x \in Q \mathbin{\mathsf{T}} 2I : \varphi(x) > \mu\}.$$

Here the summands are disjoint and open in $Q \top 2I$; according to (16) and (14), Q is contained within the first, and $Q \top [\varepsilon, 2\varepsilon]$ within the second. Thus S indeed separates $Q \top 2I$ as asserted, completing the proof of 2.19.

2.20. There exists a retraction $r : Q \top 2I \to S$ mapping Q onto S.

Proof. 2.20.1. First prove the following assertion. If

$$x_1 \top \theta_1 = x_2 \top \theta_2 \quad \text{with} \quad (x_k, \theta_k) \in Q \times 2I \tag{19}$$

and if

$$x_k \top \theta_k' \in S \quad \text{with} \quad \theta_k' \in I^\circ ,$$

then $x_1 \top \theta_1' = x_2 \top \theta_2'$.

Indeed, from (19) and $-$unicity there follows $x_2 = x_1 \top (\theta_1 - \theta_2)$ on assuming $\theta_1 \geqq \theta_2$, with $\theta_1 - \theta_2 < \varepsilon$ according to (12). Then

$$x_2 \top \theta_2' = x_1 \top (\theta_1 - \theta_2 + \theta_2') \in S \, , \quad \theta_1 - \theta_2 + \theta_2' \in 2I \; ;$$

but since also $x_1 \top \theta_1' \in S$, the unicity assertion in 2.18 yields $\theta_1' = \theta_1 - \theta_2 + \theta_2'$, and hence indeed $x_2 \top \theta_2' = x_1 \top \theta_1'$.

2.20.2. From the preceding and 2.18 it follows that one may define a map r. $r : Q \top 2I \to S$ by assigning to each $x \top \theta$ with $(x, \theta) \in Q \times 2I$ the unique point $x \top \theta' \in$ $\in S$ with θ' in I° or $2I$. From compactness of Q and S, r is continuous; and obviously r has the following properties.

$$\textit{image}\ r = S = r(Q) \, , \quad r(x) = x \quad \text{for} \quad x \in S \, .$$

This concludes the preliminaries to the final result.

2.21. Proposition. Let $\top$ be a continuous gsd system with $-$unicity on a perfectly normal space P, and let Q be a compact subset with

$$Q \cap (Q \top [\varepsilon, 3\varepsilon]) = \emptyset$$

for some $\varepsilon > 0$. Then there exists a continuous map $\sigma : P \to \mathsf{R}^1$ with

$$\sigma \mid Q = 0 \, , \quad \sigma \mid (Q \top [\varepsilon, 2\varepsilon]) = 1 \tag{20}$$

and such that the set

$$S = \{x : \sigma(x) = \tfrac{1}{2}\} \cap (Q \top [0, 2\varepsilon])$$

is a compact section, connected if Q is connected.

Proof. In the situation of 2.17 set $\mu_0 = \max_Q \varphi$, and define σ and μ by

$$\sigma(x) = \frac{\max(\mu_0, \varphi(x)) - \mu_0}{1 - \mu_0} \, , \quad \frac{\mu - \mu_0}{1 - \mu_0} = \frac{1}{2} \, .$$

Then the relations (20) follow from (16) and (14); also $\sigma(x) = \frac{1}{2}$ iff $\varphi(x) = \mu$, so that S coincides with the set constructed in (17). The remaining assertions then follow from 2.19–20, concluding the proof.

3. DYNAMICAL SYSTEMS ON LOCALLY COMPACT SPACES

Again, a different section heading might be more appropriate: Perturbation of Dynamical Systems. As explained in I,1.8, the "perturbation" of a given dynamical system T is to be interpreted as any neighbourhood of T in some topology among the dynamical systems concerned. Thus the first step is to describe a reasonable topology on the set $\mathscr{D}^T_{sl}(P)$ of all continuous lsd systems on a given topological space P; and analogously for continuous ld systems, etc. Since dynamical systems were defined as maps, the customary procedure of topologizing mapping spaces (for example in Hu, 1959, III, 9) suggests itself, suitably modified to the present case of partial maps.

3.1. Definition. Assume given a topological space P. With the set $\mathscr{D}^T_{sl}(P)$ there will be associated a standard *compact-open* topology, obtained by taking for the sub-basic open sets all subsets $V(X, Y, G)$ defined as follows:

X is compact in P, Y is compact in R^+, G is open in P, and $V(X, Y, G)$ consists of all continuous lsd systems T on P such that

$$X \times Y \subset \textit{domain}\, \mathsf{T}\,, \quad X \mathsf{T} Y \subset G\,. \tag{1}$$

For ld systems merely replace $\mathscr{D}^T_{sl}(P)$ and R^+ by $\mathscr{D}^T_l(P)$ and R^1.

3.2. Example. It is good form to justify definitions such as the preceding by reference to differential ld systems. Thus let G be an open subset in R^n, I an open interval in R^1, $f : G \times I \to \mathsf{R}^n$ continuous; and assume that, for each fixed $\lambda \in I$, the autonomous differential equation in G

$$\frac{\mathrm{d}x}{\mathrm{d}\theta} = f(x, \lambda) \tag{2}$$

has unicity of solutions. According to IV,1.3.2, each equation (2) defines a continuous lsd system T_λ. Now, the T_λ depend on λ as they should: if $\lambda \to \lambda_0$ in I, then $\mathsf{T}_\lambda \to \mathsf{T}_{\lambda_0}$ in the compact-open topology; this follows from I,1.9 and the proposition below.

3.3. Proposition. For continuous lsd systems on an LC space P, $\mathsf{T}_i \to \mathsf{T}$ in the compact-open topology is equivalent to the condition

1. If $x_j \to x$ in P, $\theta_j \to \theta$ in R^+ and $x \mathsf{T} \theta$ is defined, then

$$x_j \mathsf{T}_i \theta_j \to x \mathsf{T} \theta$$

with $x_j \mathsf{T}_i \theta_j$ defined for large i, j. (Analogously for ld systems.)

Proof. *3.3.1.* Assume $\mathsf{T}_i \to \mathsf{T}$ and the premise of 1; let G be any open neighbourhood of $x \,\mathsf{T}\, \theta$. Since T is a continuous lsd system, there exists a neighbourhood $X \times Y$ of (x, θ) in $P \times \mathsf{R}^+$ with (1); and as $P \times \mathsf{R}^+$ is LC by assumption, $X \times Y$ may even be taken compact. Thus X, Y, G are as required in 3.1, and T is in the corresponding open set $V(X, Y, G)$. From $\mathsf{T}_i \to \mathsf{T}$ there then follows $\mathsf{T}_i \in V(X, Y, G)$ for large i; and since $(x_j, \theta_j) \to (x, \theta)$, one has $(x_j, \theta_j) \in X \times Y$ for large j. Hence, for large i and j, $x_j \,\mathsf{T}_i\, \theta_j$ is defined and within G, and as G was an arbitrary open set containing $x \,\mathsf{T}\, \theta$, this proves that indeed

$$x_j \,\mathsf{T}_i\, \theta_j \to x \,\mathsf{T}\, \theta .$$

3.3.2. For the converse implication, assume 1 holds, and take any $V(X, Y, G)$ containing T. First show that each $(x, \theta) \in X \times Y$ has an open neighbourhood $U = U_1 \times U_2$ in $P \times \mathsf{R}^+$ such that, for large i,

$$U \subset \mathit{domain}\, \mathsf{T}_i\,, \quad U_1 \,\mathsf{T}_i\, U_2 \subset G\,. \tag{3}$$

In the opposite case there is a point $(x, \theta) \in X \times Y$ such that in each neighbourhood U there exist (x_U, θ_U) with

$$x_U \to x\,, \quad \theta_U \to \theta\,,$$

and either $(x_U, \theta_U) \notin \mathit{domain}\, \mathsf{T}_i$ for arbitrarily large i and small U, or $x_U \,\mathsf{T}_i\, \theta_U \notin G$ for like i and U. However, each of these contradicts 1.

3.3.3. Therefore to each $(x, \theta) \in X \times Y$ there is indeed a U as indicated, with (3) holding for all $i \geqq i(U)$, say. The sets U then constitute an open cover of the compact set $X \times Y$; let $\{U_k \mid 1 \leqq k \leqq n\}$ be an open subcover, and take $i_0 \geqq$ all $i(U_k)$. From (3) one then has that

$$X \times Y \subset \mathit{domain}\, \mathsf{T}_i\,, \quad X \,\mathsf{T}_i\, Y \subset G \quad \text{for} \quad i \geqq i_0\,.$$

By definition, $\mathsf{T}_i \in V(X, Y, G)$ for large i; and as the subbasic set was taken arbitrarily, this proves $\mathsf{T}_i \to \mathsf{T}$ in the compact-open topology, and concludes the proof of 3.3.

3.4. Remark. Proposition 3.3 may be interpreted as asserting that the point $x \,\mathsf{T}\, \theta$, if defined, depends continuously on all three of x, T, θ, in the appropriate topologies; also see remarks in IV,1.1. A reader familiar with mapping spaces will recognize in the expression $x \,\mathsf{T}\, \theta$ the (partial) evaluation map.

3.5. Corollary. Let T, T_0 *denote continuous lsd systems on an LC space,* α_x^{T} *the corresponding escape times. Then*

$$\alpha_{x_0}^{\mathsf{T}_0} \leqq \liminf_{\substack{x \to x_0 \\ \mathsf{T} \to \mathsf{T}_0}} \alpha_x^{\mathsf{T}}\,;$$

and similarly for ld systems and also for the negative escape times.

Proof. As in IV,1.4, applying 3.3. As a third variation on the proof method of 1.1, one has the

3.6. Lemma. Let T *be a continuous lsd system on an LC space* P, *and* $x \in P$. *If there exist continuous lsd systems* T_i *on* P *and* $(x_i, \theta_i) \in P \times \mathsf{R}^+$ *with* $x_i = x_i \mathsf{T}_i \theta_i$ *and*

$$x_i \to x\,, \quad 0 < \theta_i \to 0\,, \quad \mathsf{T}_i \to \mathsf{T}$$

in the corresponding topologies, then x *is a critical point of* T.

Proof. For a $\theta \in [0, \alpha_x^{\mathsf{T}})$ let n_i be the integral part of θ/θ_i; then, from the assumptions and 3.3,

$$x \leftarrow x_i = x_i \mathsf{T}_i n_i \theta_i \to x \mathsf{T} \theta\,,$$

proving that $x = x \mathsf{T} \theta$ for all $\theta \in [0, \alpha_x^{\mathsf{T}})$. Hence x is indeed a critical point of T.

3.7. Corollary. For continuous lsd systems T *on an LC space, a non-critical point of* T *remains non-critical under small perturbations of* T.

Proof. From 3.6, if L is a subset of $\mathscr{D}_{sl}^T(P)$ closed in the compact-open topology, then the set of critical points of $\mathsf{T} \in L$ is closed in P.

3.8. Lemma. If P *is an LC space, then* $\mathscr{D}_{sl}^T(P)$ *with the compact-open topology is a* T_2 *space, and* $\mathscr{D}_{sg}^T(P)$ *is a* T_3 *space; similarly for* $\mathscr{D}_l^T(P)$ *and* $\mathscr{D}_g^T(P)$.

Proof. 3.8.1. It is required to verify that distinct elements of $\mathscr{D}_{sl}^T(P)$ have disjoint neighbourhoods; the pertinent property of the sub-basic open sets is

$$V(X, Y, G_1) \cap V(X, Y, G_2) = V(X, Y, G_1 \cap G_2)\,. \tag{4}$$

Take continuous lsd systems $\mathsf{T}_1 \neq \mathsf{T}_2$ on the LC space P. Since P is a T_2 space, according to the third unicity theorem there must exist $(x, \theta) \in P \times \mathsf{R}^+$ with both $x \mathsf{T}_k \theta$ defined but $x \mathsf{T}_1 \theta \neq x \mathsf{T}_2 \theta$. Hence there exist disjoint open neighbourhoods G_k of the $x \mathsf{T}_k \theta$; and then, from (4), $V(x, \theta, G_k)$ are disjoint open neighbourhoods of the T_k.

3.8.2. Next, consider only global continuous lsd systems; then, using 3.3, it is easily shown that $\bar{G} \subset G'$ implies

$$\overline{V(X, Y, G)} \subset V(X, Y, \bar{G}) \subset V(X, Y, G')\,,$$

and regularity of $\mathscr{D}_{sl}^T(P)$ follows directly.

3.9. On T_2 carriers at least, the singleton sections are precisely the non-critical points (cf. II,6.9); thus an alternative formulation of 3.7 is that singleton sections remain sections under small perturbations. Even for compact non-singleton section, this is usually not the case, as will now be shown.

Examples. 3.10.1. Let T be our usual continuous gd system in R^2, defined in complex notation by

$$z \mathsf{T} \theta = z + \theta \quad \text{for} \quad z \in \mathsf{R}^2,\ \theta \in \mathsf{R}^1,$$

and consider the cubic parabola

$$S = \{x + iy : y = x^3,\ \ x, y \in \mathsf{R}^1\}.$$

Obviously S is a global section (cf. II,6.5 and Fig. 1). For any fixed $\varphi \in \mathsf{R}^1$ consider the isomorphic image T_φ of T under the map $f : \mathsf{R}^2 \approx \approx \mathsf{R}^2$, $f(z) = ze^{i\varphi}$; then

$$z \mathsf{T}_\varphi \theta = f(f^{-1}(z) \mathsf{T} \theta) = z + e^{i\varphi}\theta,$$

and, directly from 3.3, $\mathsf{T}_\varphi \to \mathsf{T}$ for $\varphi \to 0$ in the compact-open topology. Finally, S is a section relative to T_φ with $-\pi < \varphi \leqq \pi$ if $|\varphi - \frac{3}{4}\pi| \leqq \frac{1}{4}\pi$; in particular, even for small $\varphi > 0$, S is not a section to T_φ, and the same holds for any subarc of S with 0 as non-end point.

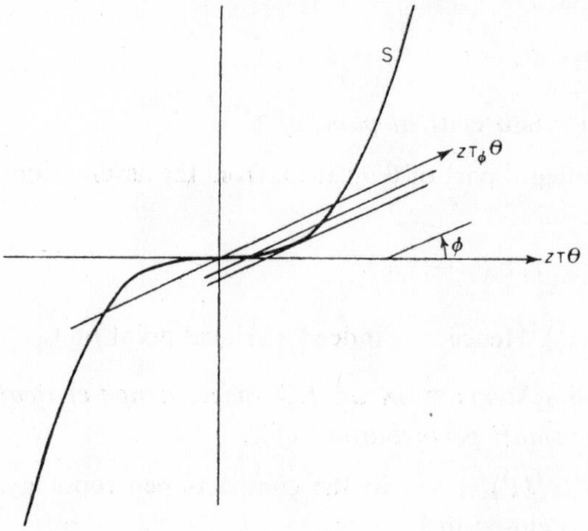

Fig. 1. Behaviour of sections under small perturbations.

3.10.2. Another phenomenon may be illustrated on a related example; let the notation of 3.10.1 be preserved. Take points $z_n = 2^{-n} + i \,.\, 2^{-3n}$ on S, and determine the slope φ_n of S at z_n, $\tan \varphi_n = 3.2^{-2n} \to 0$. Set

$$S' = (0) \cup \{z_n : n \in \mathsf{C}^+\};$$

and consider the continuous gd systems $\mathsf{T}'_n = \mathsf{T}_{\varphi_n}$. Since obviously none of the $z_m - z_k \neq 0$ has any φ_n as slope, S' is a section relative to T'_n; and the corresponding maximal t-extent is determined by the distance of z_n from 0,

$$|z_n| = 2^{-n} \sqrt{(1 + 2^{-4n})} \to 0.$$

Thus S' is a compact section relative to all T'_n and also to $\mathsf{T} = \lim \mathsf{T}'_n$, but no t-extent of S' is common for all the T'_n.

3.11. The preceding examples suggest defining *regular perturbations* of a given continuous ld system T_0 on P as subsets $L \subset \mathscr{D}_l^T(P)$ such that any point $x \in P$ non-critical relative to T_0 has a section S with the following properties: $x \in S$, $S \mathsf{T} \mathsf{T}(-\delta, \delta)$ is a neighbourhood of x for some $\delta > 0$ (cf. Section 2), for some $\varepsilon > 0$, S is a section of t-extent ε relative to all $\mathsf{T} \in L$ sufficiently near T_0. The examples 3.10 then suggest that $\mathscr{D}_l^T(P)$ may not be a regular perturbation; this can be proved using VII,2.5 for any 2-manifold P. It can also be shown that the set of differential ld systems on R^2 is a regular perturbation (by modifying the proof of VII,3.2). These questions will not be pursued further here.

This section will be concluded by presenting some specific properties of continuous dynamical systems on LC carriers.

3.12. Lemma. Let T *be a continuous lsd system on an LC space* P, *and* $x \in P$. *If* L_x^+ *is compact non-void, then for each neighbourhood* U *of* L_x^+ *and some* $\theta \in \mathsf{R}^+$,

$$x \,\mathsf{T}\, [\theta, +\infty) \subset U \,. \tag{5}$$

Proof. One need only consider U open with compact closure. Since $\emptyset \neq L_x^+ \subset U$, there exist $\theta_n \to +\infty$ in R^+ with $x \,\mathsf{T}\, \theta_n \in U$. If (5) holds for no $\theta \in \mathsf{R}^+$, there exist $\theta_n' \to +\infty$ with $x \,\mathsf{T}\, \theta_n' \notin U$; and then there must also exist $\theta_n'' \to +\infty$ with $x \,\mathsf{T}\, \theta_n'' \in$ $\in \operatorname{Fr} U$. Since $\operatorname{Fr} U \subset \overline{U}$ is compact, one may assume $x \,\mathsf{T}\, \theta_n'' \to x'$; but then

$$x' \in L_x^+ \cap \operatorname{Fr} U$$

contradicts $L_x^+ \subset U = \operatorname{Int} U$, concluding the proof of 3.12.

3.13. Corollary. For continuous lsd systems on LC spaces, a +trajectory is Lagrange +stable iff L_x^+ *is compact non-void.*

Proof. IV,4.7, and 3.12 with $\overline{U}$ taken compact.

Another by-product of 3.12 is the following proposition, implying that in certain cases *ao*+ stability implies uniform *ao*+ stability.

3.14. Proposition. Let T *be a continuous lsd system on an LC space* P, *and* $X \subset P$ *an o+ stable subset. Then every compact +invariant region of attraction for* X *is a region of uniform attraction.*

Proof. Let V be a region of attraction as indicated, and assume it is not a region of uniform attraction. Then there is a neighbourhood U of X with $x_n \,\mathsf{T}\, \theta_n \notin U$ for some $x_n \in V$, $\theta_n \to +\infty$ in R^+. From the assumptions on X (and IV,4.9, IV,2.13), U may be taken open +invariant. Since V is compact, one may assume $x_n \to x$ in V; and since it is a compact +invariant region of attraction for X,

$$\emptyset \neq L_x^+ \subset X$$

from IV,4.7; hence from 3.12, for some $\theta \in \mathsf{R}^+$ there is

$$x \,\mathsf{T}\, [\theta, +\infty) \subset U \,.$$

As U is open, one may take n large enough to have both

$$\theta + 1 < \theta_n \,, \quad x_n \,\mathsf{T}\, (\theta + 1) \in U \,.$$

But then $x_n \,\mathsf{T}\, \theta_n \in U$ from +invariance of U. This contradiction completes the proof of 3.14.

3.15. To a non-compact LC space P there is the familiar one-point-compactification $\dot{P}$ (Bourbaki, 1951, I, § 10,8); this is obtained by adjoining to P a single further point, conveniently denoted by ∞, and by taking as basic open sets in $\dot{P}$ all sets originally open in P, and also all sets of the form $(\infty) \cup (P - X)$ with X compact in P.

Now assume there is also given a continuous gd system T on P; for $\theta \in \mathsf{R}^1$ set

$$x \,\dot{\mathsf{T}}\, \theta = x \,\mathsf{T}\, \theta \quad \text{if} \quad x \in P\,, \quad \infty \,\dot{\mathsf{T}}\, \theta = \infty\,. \tag{6}$$

A similar construction may be performed for continuous gsd systems. In either case there results an abstract global system $\dot{\mathsf{T}}$ of the same type as T, and one then inquires about the continuity of $\dot{\mathsf{T}}$.

3.16. Proposition. If T is a continuous gd system on a non-compact LC space P, then there exists a unique continuous gd system $\dot{\mathsf{T}} \supset \mathsf{T}$ on $\dot{P}$, and it is described by (6). The +limit sets of T and $\dot{\mathsf{T}}$ are related thus : for $x \in P$

$$L_x^+ = \dot{L}_x^+ \cap P = \dot{L}_x^+ - (\infty)\,.$$

Proof. 3.16.1. In proving unicity, $\dot{\mathsf{T}} \supset \mathsf{T}$ implies the first relation in (6); and since T is a gd system, P is invariant in $\dot{P}$, so that $\infty = \dot{P} - P$ is an invariant singleton, i.e. a critical point (cf. II,5.6). This verifies (6) entirely.

3.16.2. To prove continuity of $\dot{\mathsf{T}}$ defined by (6), it suffices to consider the case that

$$P \ni x_i \to \infty \text{ in } \dot{P}\,, \quad \theta_i \to \theta \text{ in } \mathsf{R}^1\,. \tag{7}$$

If $x_i \,\mathsf{T}\, \theta_i \not\to \infty \,\dot{\mathsf{T}}\, \theta = \infty$, then $x_j \,\mathsf{T}\, \theta_j \to x' \in P$ for some subsequence, implying

$$x_j = (x_j \,\mathsf{T}\, \theta_j)\,\mathsf{T}\,(-\theta_j) \to x' \,\mathsf{T}\,(-\theta) \in P$$

in contradiction with $x_i \to \infty$; hence (7) indeed implies $x_i \,\mathsf{T}\, \theta_i \to \infty \,\dot{\mathsf{T}}\, \theta$. This completes the proof of 3.16.

3.17. Corollary. Each continuous gd system on R^n is isomorphic to the relativization of a continuous gd system on S^n to the open set $\mathsf{S}^n - (\infty)$.

3.18. Corollary. Each continuous gd system on a simply connected region in R^2 is isomorphic to the relativization of a continuous gd system on S^2 to the open set $\mathsf{S}^2 - (\infty)$.

Proof. A carrier as indicated is homeomorphic to R^2.

4. DYNAMICAL SYSTEMS ON MANIFOLDS

This section treats some special properties of continuous dynamical systems on topological manifolds, and on their generalization, the manifolds with boundary. One of the most elegant of these is

4.1. Proposition. Each continuous lsd system with $-$unicity on a manifold P has the openness property, and thus is the semi-dynamical restriction of a unique continuous ld system on P.

Proof. To recall the terms employed, see IV,2.4 and II,2.3. Let T be the lsd system as indicated, and take any open $G \subset P$ and $\theta \in \mathsf{R}^+$. Then

$$G \mathsf{T} \theta = H \mathsf{T} \theta \quad \text{with} \quad H = \{x \in G : x \mathsf{T} \theta \text{ defined}\}\,; \tag{1}$$

and directly from the definition IV,1.1.2, H is open in P. From V,2.8, the translation T_θ defines a homeomorphism $H \approx H \mathsf{T} \theta$, whereupon $H \mathsf{T} \theta$ is open in P according to the preservation of domain theorem. Hence in (1), $G \mathsf{T} \theta$ is open, so that T has the openness property. Finally apply IV, 2.5 and a local determinacy theorem.

4.2. Theorem. Every continuous lsd system with $-$unicity on a compact manifold P is global, and is the semi-dynamical restriction of a unique continuous gd system on P.

Proof. 4.1, IV,1.10.

Next, consider manifolds with boundary; as shown by the natural gsd system on R^+, proposition 4.1 cannot be carried over to these carriers. However, the openness property is connected with the absence of *s*-points (cf. IV,2.5.2 and II,2.9), and from 4.1 absence of boundary points implies openness. It is then not surprising that there is an intimate connection between *s*-points and boundary points. This is exhibited in the following result; also 4.4 may be compared with $-$invariance of the *s*-points set, II,5.6.

4.3. Lemma. If T is a continuous lsd system with $-$unicity on a manifold with boundary P, then each s-point of T is on the boundary of P.

Proof. It suffices to show that T has no *s*-points in any component P' of the set P_0 of non-boundary points of P. Since P is locally connected and P_0 open, P' is open connected (Kuratowski, 1950, II, § 44, II, 4); evidently P' has no boundary points, so that it is a manifold. On P', the system T relativizes to a continuous lsd system T' with $-$unicity (cf. IV,2.10 and II,2.7).

Now apply 4.1 to T' on P'; thus T' is the restriction of an ld system on P', and in particular T' has no *s*-points in P' (cf. II,2.9). Since P' is open in P, it follows that T also has no *s*-points in P' as asserted. This concludes the proof.

4.4 Lemma. If T is a continuous lsd system with $-$unicity on an n-manifold with boundary P, then the boundary of P is $-$invariant.

Proof. An equivalent formulation, of course, is that the set of non-boundary points is $+$invariant. Thus, take a non-boundary point $x \in P$, with $x \mathsf{T} \theta$ defined for some $\theta \in \mathsf{R}^+$. By definition, $x \mathsf{T} \theta$ has a neighbourhood $U \approx \mathsf{E}^n$, and x has a neighbourhood $V \approx \mathsf{R}^n$ mapped homeomorphically into U by T_θ (cf. V,2.8); and it remains to prove that $x \,\mathsf{T}\, \theta = T_\theta x$ is a non-boundary point of U. Inject U into R^n via the homeomorphism $U \approx \mathsf{E}^n$; the preservation of domain theorem yields that $T_\theta V$ is open in R^n, and hence $T_\theta V \subset U$ implies $T_\theta V \subset \operatorname{Int} U$ (interior in R^n, of course). This concludes the proof.

4.5. Corollary. If T *is a continuous ld system on a manifold with boundary P, then the boundary of P is invariant.*

Proof. 4.4 and orientation change.

Next, some results will be given concerning the section-extension procedure described in 2.12, in connection with manifolds as carrier spaces. First it will be shown that the requirement from 2.1 on the extended section, namely that $S' \,\mathsf{T}\, (-\delta, \delta)$ is to be a neighbourhood of S, corresponds to the condition that the dimension of S' be as large as possible.

4.6. Lemma. Let T *be a continuous ld system on an n-manifold P, and let a manifold with boundary* $S \subset P$ *be a section of t-extent* ε *and such that* $S \times I \subset$ *domain* T *for* $I = [-\frac{1}{2}\varepsilon, \frac{1}{2}\varepsilon]$. *Then* $\dim S \leqq n - 1$, *with equality iff* $\operatorname{Int}(S \,\mathsf{T}\, I) \neq \emptyset$.

Proof. 4.6.1. From IV,2.17, $P \supset S \,\mathsf{T}\, I \approx S \times I$; since S is a manifold with boundary, this implies

$$n = \dim P \geqq \dim (S \,\mathsf{T}\, I) = \dim (S \times I) = \dim S + 1\,,$$

so that in any case one must have $\dim S \leqq n - 1$.

4.6.2. If $\operatorname{Int}(S \,\mathsf{T}\, I) \neq \emptyset$, then $S \,\mathsf{T}\, I$ contains a set open in the n-manifold P, and hence also a set homeomorphic to R^n; thus

$$n \leqq \dim (S \,\mathsf{T}\, I) = \dim S + 1\,.$$

With the previous result, this yields $\dim S = n - 1$.

Conversely, if $\dim S = n - 1$, then $S \,\mathsf{T}\, I \approx S \times I$ is an $(n - 1 + 1)$-manifold with boundary, and hence contains a set U homeomorphic to R^n; from the preservation of domain theorem, U is open in the n-manifold P. This concludes the proof of 4.6.

In the section-extension procedure it was obtained that the extended section S' has

$$\operatorname{Int}(S' \,\mathsf{T}\, [-\delta, \delta]) \supset \operatorname{Int}(S' \,\mathsf{T}\, (-\delta, \delta)) \supset S\,,$$

and thus 4.6 may be applied if the original section S was non-void. Next we will consider the question whether or not $S' \neq S$.

4.7. Lemma. Every compact $(n - 1)$*-manifold section S of a continuous ld system* T *on an n-manifold is inextensible in the sense that each connected section* $S' \supset S$ *coincides with S.*

Proof. 4.7.1. Take any section S' as indicated, with t-extent say ε. For small positive $\delta < \frac{1}{2}\varepsilon$ it follows from IV,2.17 that $S \top (-\delta, \delta) \approx S \times (-\delta, \delta)$, so that $G = S \top (-\delta, \delta)$ is an n-manifold. Hence by the preservation of domain theorem, G is an open set containing S.

4.7.2. Now prove that

$$S' \cap G = S.$$

One inclusion follows from $S' \supset S \subset G$. For the converse, note that any $x' \in S' \cap G$ has $S' \ni x' = x \top \theta$ with $x \in S \subset S'$ and $|\theta| < \delta \leqq \frac{1}{2}\varepsilon$; as S' is a section of t-extent ε, necessarily $x' = x \in S$.

4.7.3. From 4.7.2,

$$S' = (S' \cap G) \cup (S' - G) = S \cup (S' - G);$$

where S is compact, G open and S' connected, there follows $S' - G = \emptyset$ and hence $S' = S$ as asserted. (If S were empty, then the carrier would be a 0-dimensional manifold, i.e. a singleton critical point, and then all sections are void sets.) This concludes the proof.

4.8. Proposition. In a continuous ld system $\top$ *on an n-manifold P, let* $S \subset S'$ *be compact sections and manifolds with boundary such that* $S' \top (-\varepsilon, \varepsilon)$ *is a neighbourhood of S for some* $\varepsilon > 0$ *with*

$$2\varepsilon \leqq t\text{-extent of } S', \quad S' \top (-\varepsilon, \varepsilon) \subset \text{domain } \top.$$

Then $S = S'$ *iff S is an* $(n-1)$*-manifold.*

Proof. It may be assumed that S is non-void. In any case, from the assumptions and 4.6 one has that $\dim S' = n - 1$. Having 4.7, it remains to prove $S \neq S'$ if S is not an $(n-1)$-manifold. There are two cases; either the dimension of S is less than $n - 1 = \dim S'$ and then, of course, $S \neq S'$; or the dimension of S is the maximal allowed by $S \subset S'$ but S contains a boundary point, x. From IV,2.17, $S' \top (-\varepsilon, \varepsilon) \approx$ $\approx S' \times (-\varepsilon, \varepsilon)$ and is therefore an n-manifold with boundary; this boundary is obviously generated by that of S'. Now S is within the open set $\operatorname{Int}(S' \top (-\varepsilon, \varepsilon))$, so that no point of S is a boundary point of S'. In particular, the boundary point x of S is non-boundary in S', and one again concludes that $S \neq S'$. This completes the proof of 4.8.

4.9. The last question treated in this chapter is the classification problem for continuous lsd systems with $-$unicity on 1-manifolds. In the course of this it is shown that every continuous ld system on an 1-manifold is isomorphic to a differential ld system. Some intermediate steps are separated out as lemmas.

4.10. Lemma. Every continuous gd system $\top$ *on* R^1 *without critical points is isomorphic to the natural gd system.*

Proof. Since R^1 contains no simple closed curve, from the assumptions and 1.6 it follows that every trajectory in R^1 is an open-interval. Now, distinct trajectories are always disjoint and R^1 is connected, so that R^1 is a single trajectory. Therefore $t_x : \mathsf{R}^1 \approx \mathsf{R}^1$ for any chosen solution t_x of T; and $t_x : (\mathsf{R}^1, +) \to (\mathsf{R}^1, \mathsf{T})$ is a morphism in $\mathscr{D}_g$:

$$t_x(x' + \theta) = x \,\mathsf{T}\, (x' + \theta) = (x \,\mathsf{T}\, x')\, \mathsf{T}\, \theta = t_x(x')\, \mathsf{T}\, \theta\,.$$

Thus t_x is the required isomorphism in $\mathscr{D}_g^T$, concluding the proof.

4.11. Lemma. Every continuous gd system T *on* R^1 *is isomorphic to a differential gd system.*

Proof. 4.11.1. Having the previous result, it may be assumed that the critical point set K of T is non-void. According to 1.2, the set $\mathsf{R}^1 - K$ is open, and hence decomposes into a countable system of contiguous intervals $J_n : \mathsf{R}^1 - K = \bigcup J_n$. Within each J_n choose a point x_n, and denote by t_n the solution of T through x_n,

$$t_n : \mathsf{R}^1 \approx J_n\,, \quad t_n(\theta) = x_n \,\mathsf{T}\, \theta_n\,.$$

4.11.2. Now define a differential gd system on R^1 as follows. For $x \in J_n$ let $\sigma(x) = 1$ or -1 according to whether t_n increases or decreases, for $x \in K$ set $\sigma(x) = 0$. Secondly, let $\psi(x)$ be the Euclidean distance of $x \in \mathsf{R}^1$ from K, so that $K \neq \emptyset$ implies $\psi : \mathsf{R}^1 \to \mathsf{R}^+$ is continuous; and set

$$\varphi(x) = \sigma(x) \exp - \psi^{-2}(x)\,, \tag{2}$$

(see Fig. 2). Obviously φ is of class C^∞ and bounded by $1 \,.\, e^\circ = 1$; hence (i.e. from IV,1.3.2, I,1.6, I,1.11) the differential equation $dx/d\theta = \varphi(x)$ defines a differential gd system T' in R^1.

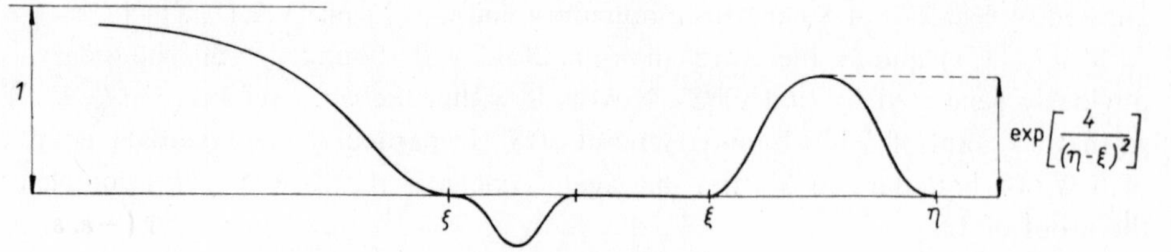

Fig. 2. Graph of φ in contiguous intervals $(-\infty, \xi)$ and (ξ, η).

4.11.3. According to (2), the system T' has the same critical point set K as T. Again let t'_n be the solution of T' through the previously chosen point x_n, so that

$$t'_n(\theta) = x_n \,\mathsf{T}\, \theta\,;$$

and one has the diagram

$$\begin{array}{ccc} t_n : \mathsf{R}^1 & \approx & J_n \\ \Vert & & \uparrow f_n \\ t'_n : \mathsf{R}^1 & \approx & J_n \end{array}$$

where f_n is defined by requiring commutativity, i.e. $f_n = t_n t_n'^{-1}$. From (2) and the definition of σ, the f_n is an increasing homeomorphism $J_n \approx J_n$.

4.11.4. Finally, define a map $f : \mathsf{R}^1 \to \mathsf{R}^1$ by the following assignments:

$$f(x) = x \quad \text{if} \quad x \in K\,, \quad f \mid J_n = f_n$$

(for all J_n). From the preceding construction, f maps onto R^1 and strictly increases; hence it is continuous, and therefore a homeomorphism. Now verify that f is a morphism in $\mathscr{D}_g$; take any $x \in \mathsf{R}^1$, $\theta \in \mathsf{R}^1$. If $x \in K$, then

$$f(x \mathsf{T}' \theta) = f(x) = x = x \mathsf{T} \theta = f(x) \mathsf{T} \theta$$

as required. If x is in some J_n, then $x = x_n \mathsf{T}' \theta'$ for a $\theta' \in \mathsf{R}^1$, whereupon

$$\begin{aligned} f(x \mathsf{T}' \theta) &= f(x_n \mathsf{T}' (\theta' + \theta)) = t_n t_n'^{-1}(x_n \mathsf{T}' (\theta' + \theta)) \\ &= t_n(\theta' + \theta) = x_n \mathsf{T} (\theta' + \theta) = (x_n \mathsf{T} \theta') \mathsf{T} \theta\,; \end{aligned} \tag{3}$$

in particular $f(x) = x_n \mathsf{T} \theta'$ on setting $\theta = 0$, and hence in (3)

$$f(x \mathsf{T}' \theta) = (x_n \mathsf{T} \theta') \mathsf{T} \theta = f(x) \mathsf{T} \theta\,.$$

Therefore indeed

$$f : (\mathsf{R}^1, \mathsf{T}') \approx (\mathsf{R}^1, \mathsf{T}) \text{ in } \mathscr{D}_g^T\,,$$

concluding the proof of 4.11.

4.12. From the preceding proof it also follows that each continuous gd system on R^1 is characterized, up to isomorphism in $\mathscr{D}_g^T$, by a closed subset $K \subset \mathsf{R}^1$ and by an assignment of ± 1 to each contiguous interval of $\mathsf{R}^1 - K$ (the critical point set and the orientation of trajectories, respectively). The differential gd system defined in 4.11.2 might be termed the *canonic representation* of the given system T.

4.13. Lemma. Each continuous ld system T *on* R^1 *is isomorphic to a differential ld system.*

Proof. 4.13.1. Having 4.11, it may be assumed that T is non-global. First show that the escape times α_x of T depend continuously on x. At a critical point x, continuity follows from $\alpha_x = +\infty$ and lower semi-continuity, IV,1.4. If x is non-critical, then the solution $t = t_x$ of T through x is a local homeomorphism onto a neighbourhood of $x = t(0)$, 1.6; and continuous dependence of $\alpha_{t(\theta)} = \alpha_{x \mathsf{T} \theta} = \alpha_x - \theta$ on θ follows from II,1.5.

4.13.2. Hence, according to V,3.8.1, the maximal gd extension $\hat{\mathrm{T}}$ of T has as carrier a T_2 space; from V,3.5 then, $(\mathrm{R}^1)^\wedge$ is a 1-manifold. Now, there are precisely two 1-manifolds, R^1 and S^1; and $(\mathrm{R}^1)^\wedge$ is non-compact according to V,3.3, since T was assumed non-global. Thus $(\mathrm{R}^1)^\wedge \approx \mathrm{R}^1$, $\hat{\mathrm{T}}$ is a continuous gd system on R^1, and therefore isomorphic to a differential gd system T'. Finally T, the relativization of $\hat{\mathrm{T}}$ to a connected open subset (cf. V,3.1), is isomorphic to the relativization of the differential system T' over the corresponding open-interval in R^1. This concludes the proof of 4.13.

4.14. Theorem. Every continuous ld system on a 1-manifold is isomorphic to a differential ld system.

Proof. 4.14.1. As already mentioned, there are only two 1-manifolds, R^1 and S^1, and the former of these has been treated in 4.13. Thus, let T be a continuous ld system on S^1, necessarily global. According to IV,3.13. the projection map

$$p : \mathrm{R}^1 \to \mathrm{S}^1 , \quad p(x) = \exp ix/(2\pi) , \tag{4}$$

representing R^1 as a covering space of S^1, induces a "covering" gd system T' on R^1 with

$$p : (\mathrm{R}^1, \mathrm{T}') \to (\mathrm{S}^1, \mathrm{T})$$

a morphism in $\mathscr{D}_g^T$.

4.14.2. It follows that T' is invariant relative to F under the simple translation

$$F : \mathrm{R}^1 \approx \mathrm{R}^1 , \quad F(x) = x + 1 ;$$

in particular, $F(x)$ is a critical point iff x is critical. It is readily verified that the canonic representation T'' of T' is then also F-invariant. Since the exponential map (4) is of class C^∞, T'' induces a differential gd system on S^1, obviously isomorphic with the original T. This completes the proof.

4.15. Finally, consider the classification of continuous lsd systems T with $-$unicity on 1-manifolds with boundary; these latter are

$$\mathrm{R}^1, \mathrm{S}^1, \mathrm{R}^+, \mathrm{E}^1 .$$

The first two of these are 1-manifolds; thus according to 4.1 and 4.14, the given T is the lsd restriction of a continuous ld system, isomorphic to a differential ld system. As for the last two carriers, their boundary is $-$invariant, and hence each boundary component, a singleton, is critical or an s-point (4.4 and II,5.6). On the remaining $+$invariant portion, a 1-manifold, the given T relativizes to a continuous lsd system with $-$unicity, and may then be processed as in the former two cases.

Possibly it is recognized that the proofs of 4.10–15 are merely an exercise in the application of previous, and considerably deeper, results. It seems that any attempt to extend 4.14 to higher dimensions would have to employ principles more refined than the brute force methods used, for example, in 4.11. This is also suggested by the related but far more restricted result of VIII,4.6.

CHAPTER VII

Transversal Theory

In the following chapter, the classical results on limit sets of differential ld systems on R^2 are carried over to the case of continuous ld systems on 2-manifolds. The purpose of the present chapter is to develop the necessary apparatus, an abstract variant of classical transversal theory. To this end it was necessary, first, to select, among several reasonable candidates, an adequate analogue of the concept of differential transversals. Then topological proofs had to be devised to replace the analytic ones; once found, these usually proved considerably simpler and occasionally more general than the original ones.

In the situation of differential ld systems, there are no problems as to the existence of sufficiently many transversals, since one has available the isogonals to trajectories, or even the rectilinear normals (cf. for example, Coddington and Levinson, 1955, XVI,2.1a). For continuous ld systems, the Whitney-Bebutov theorem VI,2.13 applies; this yields sufficiently many sections with reasonable properties, and Section 1 is devoted to showing that many of these are curves. (Theorem 1.6 was obtained in Hájek 5, 1965 and 6, 1965.) This forms the background of the definition of transversals adopted in 2.1.

Canonic domains (the characteristic rectangles of Lefschetz, 1951, IX, § 1) are introduced in Section 2, together with the related canonic homeomorphisms. The latter make it possible to replace heuristic proofs within characteristic rectangles on the carrier by precise but no less vivid proofs within geometrical rectangles in the canonic images. The mean-value theorem 2.16 may be compared with Lemma 2.1(b) in (Coddington and Levinson, 1955, XVI).

In Section 4 there is presented a series of theorems concerning the configurations formed by intersecting trajectories and transversals. The non-intersection theorems of Duff (1953, § 2), theorem 2 of Seifert (1958) and theorem I of Jel'šin (1954) are all related to theorem 4.4; however, they concern the situation of two differential gd systems with mutually transversal trajectories rather than a single transversal and trajectory. The three-point theorem 4.10 is also familiar for differential ld systems: for example, Coddington and Levinson (1955, XVI,2.2) with rectilinear transversals, Lefschetz (1957, X, § 3, 6.2) for a heuristic argument. The method adopted in the present chapter in proving these theorems via the canonic map theorem 2.5 was developed in Hájek 1 (1960) for differential gd systems in R^2, essentially by proving 3.2; this includes the mean-value theorem, and the monotonicity theorem in the formulation of 4.14. The presentation below is taken over, with minor modifications, from Hájek 8 (1965, §§ 3–4).

The standing assumption in the present chapter is that P is a 2-manifold and T a continuous ld system on P; this will not be repeated except, for emphasis, in the principal assertions.

1. CONTINUA-SECTIONS ARE CURVES

In addition to the T and P specified previously, in this section there is assumed given a subset $Q \subset P$ which is both a continuum and a section with t-extent $\varepsilon > 0$.

For definiteness, a *continuum* is a compact connected set containing at least two distinct points; sections were defined in II,6.5. It will be useful to reserve the term curve to sets which, in current usage, are either simple arcs or simple closed curves. More precisely, *curves* are homeomorphs of either E^1 or S^1, i.e. the compact 1-manifolds with boundary; closed curves are homeomorphs of S^1 (with "closed curve" used as a non-divisible term). With this clarified, the contents of the present section are summarized in the heading; the proof of this assertion is divided into separately labelled steps.

1.1. The first of these steps is found ready to hand in general topology, namely the concept and properties of the so-called dendrites. In a metrisable space, a locally connected continuum is termed a *dendrite* if it contains no simple closed curves (Kuratowski, 1950, § 46, VI). The following properties are pertinent.

1.1.1. Distinct points x, y of a dendrite D can be connected by precisely one arc $\widehat{xy} \subset D$ (*loc. cit.*, 2).

1.1.2. A locally connected continuum C is a local dendrite (i.e. each $x \in C$ has a neighbourhood U with $C \cap U$ a dendrite) if the diameters of simple closed curves within C have a positive lower bound (Kuratowski, 1950, § 46, VII,3).

Since C in 1.1.2 is compact, the requirement is that each point of C have a neighbourhood containing no simple closed curves of C. Now return to the situation discribed above.

1.2. Lemma. If Q is locally connected, it is a local dendrite.

Proof. Take any $x \in Q$; there exists a $\lambda \in \mathsf{R}^1$ and a neighbourhood V of x with

$$0 < \lambda \leqq \tfrac{1}{2}\varepsilon, \quad V \cap (V \mathsf{T} \lambda) = \emptyset .$$

Indeed, if this were not the case, to each such λ one would have points $x_i \in P$ with

$$x_i \to x, \quad x_i \mathsf{T} \lambda \to x,$$

and hence $x \mathsf{T} \lambda = x$: a contradiction, since x is on a section Q and thus is non-critical (cf. II,6.9).

Next, take any disc neighbourhood U (i.e. a homeomorph of R^2) of x with $U \subset V$, and show that there is no simple closed curve $C \subset Q \cap U$. Assume the contrary; then C separates $U \approx \mathsf{R}^2$ into two regions of which precisely one, G, has compact closure in U (G corresponds to the inner region of $\mathsf{R}^2 - C$). Choosing $x' \in G \subset U \subset V$ one has $x' \mathsf{T} \lambda \notin V \supset G$, so that there must exist a $\theta \in \mathsf{R}^1$ with

$$0 < \theta < \lambda \,, \quad x' \mathsf{T} \theta \in C \,.$$

By symmetry, there is a $\theta' \in \mathsf{R}^1$ with

$$-\lambda < \theta' < 0 \,, \quad x' \mathsf{T} \theta' \in C \,;$$

with $C \subset Q$, $\lambda \leqq \frac{1}{2}\varepsilon$, this contradicts the assumption that Q is a section of t-extent ε, II,6.8.1. The proof is then concluded by applying 1.1.2.

1.3. Lemma. If Q is locally connected, then it is locally an arc.

Proof. 1.3.1. From 1.2, 1.1.1 – 2 it follows that each $x \in Q$ has a neighbourhood U such that $Q \cap U$ contains an arc, either of the form $\widehat{axb}$, or, if not, then of the form $\widehat{ax}$. In each case we proceed to prove that $Q \cap U$ coincides with the indicated arc, at least locally at x.

1.3.2. Take a continuous parametrization of $\widehat{axb}$,

$$p : [-1, 1] \approx \widehat{axb} \,, \quad p(0) = x \,, \tag{1}$$

and define a continuous map h by

$$h(\theta, \sigma) = p(\sigma) \mathsf{T} \theta \quad \text{for} \quad |\theta| \leqq \tfrac{1}{2}\varepsilon, \ |\sigma| \leqq 1 \,.$$

Since $\widehat{axb} \subset Q$ is itself a section of t-extent ε, it follows from II,6.8.3 and (1) that h is 1 – 1. Hence h is a homeomorphism, and from the preservation of domain theorem $V = \textit{image}\, h$ is a neighbourhood of x; indeed, $[-\frac{1}{2}\varepsilon, \frac{1}{2}\varepsilon] \times [-1, 1]$ is a neighbourhood of $(0, 0)$, and $h(0, 0) = p(0) \mathsf{T} 0 = x$. By choice of ε as a t-extent of Q, the set Q meets V only in the arc $\widehat{axb}$. Summarizing, V is a neighbourhood of x and $Q \cap V = \widehat{axb}$.

1.3.3. Now assume that there is no subarc $\widehat{axb}$ in $Q \cap U$, and consider the set A of all arcs

$$\widehat{ax} \subset Q \cap U \quad \text{with} \quad a \in Q \cap U \,;$$

by assumption, A is non-void. Using 1.1.1 and 1.3.2 it is easily shown that A is monotone; using compactness, the set-union of members of A is shown to belong to A again. Hence

$$Q \cap U = \widehat{ax} \quad \text{for some} \quad a \in Q \cap U \,.$$

This completes the proof of 1.3.

1.14. Lemma. If Q is locally connected then it is a curve.

Proof. This follows from 1.3, since obviously a continuum which is locally an arc at each of its points is necessarily a curve.

1.5. The final step consists in lifting the local connectedness requirement in 1.4. Observe that T, P and Q satisfy the assumptions of theorem VI,2.12, so that there exists an extended section $Q' \supset Q$; since VI,2.11 applies, one may even take Q' compact, connected and locally connected. Now Q contains at least two points, so that $Q' \supset Q$ also does; thus one may apply 1.4 to Q', obtaining that Q' is a curve. Then Q, a subcontinuum of a curve, is itself a curve. Thus we have proved

1.6. Theorem. If T *is a continuous ld system on a 2-manifold P, then each continuum-section in (P, T) is a curve.*

It may be interesting to note that the assumption that the dimension of the carrier manifold is precisely 2 came in at two points only, in 1.3.1 and in the very last sentence of 1.5.

2. TRANSVERSALS, CANONIC HOMEOMORPHISMS

Theorem 1.6 suggests that more interest might be devoted to the objects defined in 2.1; the term to be used is borrowed from differential equation theory.

2.1. Definition. A *transversal* is a curve in P which is simultaneously a section. A subset of P which is a transversal locally at an $x \in P$ is said to be *transversal at x*.

A direct consequence of IV,2.16 is that a curve in P is a transversal iff it is transversal at all of its points.

2.2. Assume given a curve C in P. Since C is compact, from IV,2.1 there follows

$$C \times \left[-\tfrac{1}{2}\delta, \tfrac{1}{2}\delta\right] \subset \mathit{domain}\ \mathsf{T} \tag{1}$$

for some $\delta > 0$; and since C is a curve, there exists a parametrization

$$p : A \approx C$$

where either $A = \mathsf{E}^1$ or $A = \mathsf{S}^1$ (according to whether C is or is not a simple arc). Also, set

$$A_\varepsilon = \left[-\tfrac{1}{2}\varepsilon, \tfrac{1}{2}\varepsilon\right] \times A \quad \text{for} \quad 0 < \varepsilon \leqq \delta\,, \tag{2}$$

and define a map $f : A_\varepsilon \to P$ by

$$f(\theta, \sigma) = p(\sigma)\, \mathsf{T}\, \theta\,; \tag{3}$$

according to (1), this is indeed possible. Having prepared this notation, transversality of C may be characterized in terms of the parametrization p:

2.3. *Lemma. C is a transversal with t-extent* $\varepsilon \leqq \delta$ *iff f is a homeomorphism; in the positive case*

$$image\, f = f(A_\varepsilon) = C\,\mathsf{T}\left[-\tfrac{1}{2}\varepsilon, \tfrac{1}{2}\varepsilon\right]$$

is a neighbourhood of all non-end points of C.

Proof. The first assertion is a special case of IV,2.17. For the second, take any non-end point x of C. Then there is a neighbourhood G of x in C with $G \approx \mathsf{R}^1$; according to the preservation of domain theorem, the homeomorphism f maps the 2-manifold $(-\lambda, \lambda) \times G$ (with $0 < \lambda \leqq \varepsilon$) onto an open set containing $f(x)$ and contained within $f(A_\varepsilon)$. This concludes the proof of 2.3.

2.4. This established, one may define a class of mappings to be called canonic homeomorphisms; these will be used extensively as apparatus in proofs and definitions. Preserving the notation of 2.2, first identify A, the domain of the parametrization p, either with some segment $[\alpha_1, \alpha_2] \subset \mathsf{R}^1$ or with the unit circle $K = \{z \in \mathsf{R}^2 : : |z| = 1\}$ in R^2; then define a map $g : A_\varepsilon \to \mathsf{R}^2$ by

$$g(\theta, \sigma) = \begin{cases} \theta + i\sigma & \text{if} \quad A = [\alpha_1, \alpha_2] \\ e^\theta \sigma & \text{if} \quad A = K\,. \end{cases} \tag{4}$$

Obviously g is a homeomorphism of A_ε into R^2, taking A_ε onto either

$$\left[-\tfrac{1}{2}\varepsilon, \tfrac{1}{2}\varepsilon\right] \times [\alpha_1, \alpha_2] \quad \text{or} \quad \{z \in \mathsf{R}^2 : |\lg|z|\,| \leqq \tfrac{1}{2}\varepsilon\}$$

(see Fig. 1).

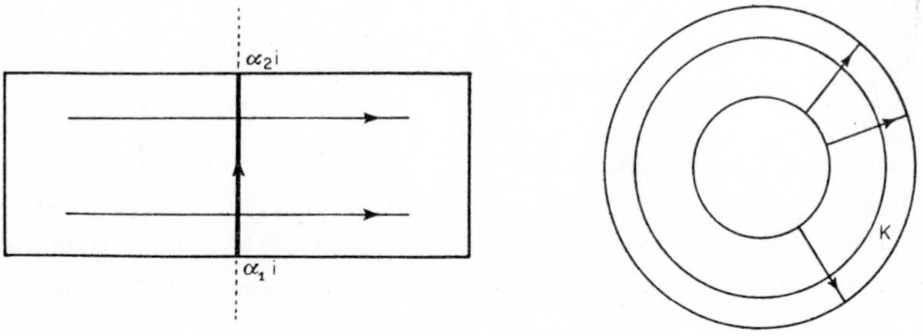

Fig. 1. Images of canonic domains.

Now assume that C is a transversal of t-extent $\varepsilon \leqq \delta$ (i.e. that f is homeomorphic, 2.3), and consider (2)–(4). The *canonic homeomorphism* h corresponding to p (and to T, ε) is defined as the map $h = gf^{-1}$. Its domain is

$$domain\, h = image\, f = C\,\mathsf{T}\left[-\tfrac{1}{2}\varepsilon, \tfrac{1}{2}\varepsilon\right],$$

and will be termed a *canonic domain*; any subset of domain h is said to be mapped canonically.

The homeomorphism h maps C and arcs of trajectories with in the canonic domain in a reasonable manner; this may be read off from the preceding or from Fig. 1. Formulating, one has

2.5. *Proposition. Let C be a transversal with t-extent ε, subject only to the condition that, for $I = (-\frac{1}{2}\varepsilon, \frac{1}{2}\varepsilon) \subset \mathsf{R}^1$, $C \times \bar{I} \subset$ domain T; also let C° be C less its end-points. Then there exists a (canonic) homeomorphism h taking $C^\circ \mathsf{T} I$ into R^2 and with the following property: h is an isomorphism in $\mathscr{D}_l^T$ taking T relativized to $C^\circ \mathsf{T} I$ into the differential ld system in R^2 defined by*

$$\frac{dz}{d\theta} = 1 \quad or \quad \frac{dz}{d\theta} = z$$

relativized to

$$I \times (-1, 1) \quad or \quad \{z \in \mathsf{R}^2 : |\lg|z|| < \tfrac{1}{2}\varepsilon\}$$

(according to whether $C \approx \mathsf{E}^1$ or $C \approx \mathsf{S}^1$); C° is mapped onto

$$\{\lambda i : \lambda \in \mathsf{R}^1,\ |\lambda| < 1\} \quad or \quad \{z \in \mathsf{R}^2 : |z| = 1\}$$

respectively; and, for all $\lambda \in (0, \frac{1}{2}\varepsilon]$, $C^\circ \mathsf{T} (-\lambda, \lambda)$ is an open set containing C°.

The Whitney-Bebutov theorem then yields as special case

2.6. *Corollary. Each non-critical point $x \in P$ has neighbourhoods which can be mapped canonically. More precisely, there exist parametrizations $p : [-1, 1] \to \to P$ of transversals such that $p(0) = x$ and that, for sufficiently small positive σ and λ, $p(-\sigma, \sigma) \mathsf{T} (-\lambda, \lambda)$ is an open canonic domain containing x.*

Proof. VI,2.12–13, 1.6, 2.5.

2.7. Occasionally, some modifications of 2.5 are useful. Note that if C is a closed curve, then $C^\circ = C$ so that $C \mathsf{T} I$ is a neighbourhood of C itself (also see VI,4.8); an analogous result may be needed in the case that $C \approx \mathsf{E}^1$. This is obtained easily by the section-extension procedure of VI,2, at the cost of decreasing the given t-extent.

In the situation of 2.4, first apply VI,2.12, obtaining a section $C' \supset C$; from VI,2.11 and 1.6, C' is again a transversal and, from VI,4.8, all points of C are non-end points in C'. Then construct a parametrization $p' \supset p$ of C', obtaining the canonic homeomorphism h'. Summarizing,

2.8. *Lemma. Let C be a simple arc transversal. Then there exist positive ε, δ and a (canonic) homeomorphism h' mapping an open neighbourhood of C into R^2, and such that h' is an isomorphism in $\mathscr{D}_l^T$, taking T relativized to domain h' into the differential ld system in R^2 defined by $dz/d\theta = 1$ relativized to $(-\varepsilon, \varepsilon) \times \times (-\delta, \delta)$; and C is mapped into the imaginary axis.*

In other cases it is advantageous to increase the t-extent as far as possible at the cost of abbreviating the transversal. This may be described as attempting to obtain a canonic domain along a trajectorial arc rather than along a transversal as in 2.8. The basis for this is indicated in the following lemma.

2.9. Lemma. Let $p : [-1, 1] \approx C$ *be the parametrization of a transversal, and let* τ *be the primitive period of the trajectory through* $p(0)$ $(\tau = +\infty$ *if this trajectory is non-cyclic). Then to any positive* $\varepsilon < \tau$ *there is a* $\sigma > 0$ *such that* $p[-\sigma, \sigma]$ *is a transversal with t-extent* ε, *and with* $p[-\sigma, \sigma] \times (-\frac{1}{2}\varepsilon, \frac{1}{2}\varepsilon) \subset$ $\subset$ *domain* T.

Proof. The last assertion follows from IV,2.1 and $p(0) \times [-\tau, \tau] \subset$ *domain* T. For the first assertion, assume the contrary. Then there exist real ε, θ_n, σ_n, σ'_n such that

$$0 < \theta_n \leqq \varepsilon < \tau\,, \quad \sigma_n \to 0\,, \quad \sigma'_n \to 0\,,$$

$$p(\sigma_n) = p(\sigma'_n)\,\mathsf{T}\,\theta_n\,;$$

furthermore, the θ_n are bounded away from 0, since C is a transversal of some positive t-extent. Thus one may assume $\theta_n \to \theta \in (0, \tau)$ and, on taking limits,

$$p(0) = p(0)\,\mathsf{T}\,\theta\,.$$

However, this contradicts the assumption on τ.

2.10. In the concluding part of this section there is exhibited a second important application of canonic homeomorphisms. If the carrier manifold can be oriented, these homeomorphisms bring into relation the possible orientations of transversals with that of the carrier, and thereby serve to distinguish between different parametrizations of the transversal curves.

2.11. Definition. Let C be an oriented transversal of a continuous ld system on an oriented 2-manifold; take any orientation-preserving parametrization p of C. Then C is termed a *+transversal* or a *−transversal* according to whether the canonic homeomorphism corresponding to p is orientation preserving or reversing.

It is, of course, implicitly understood that the natural orientation is taken for R^2 into which all canonic homeomorphism map. It may be observed that in 2.4, the homeomorphism g is orientation-preserving. Obviously the concept defined in 2.11 is independent of the particular choice of orientation-preserving parametrization used to describe C. The following assertions are immediate consequences.

2.12. Lemma. Let C *be a +transversal in* (P, T) *(with* T *a continuous ld system on an oriented 2-manifold* P*). Then*

1. $-C$ *is a −transversal in* (P, T).
2. C *is a −transversal in* $(-P, \mathsf{T})$.

3. *C is a −transversal in* (P, T^C), *with* T^C *the ld system obtained from* T *by the orientation-changing functor.*

4. fC *is a +transversal in* (P', T') *if* $f : (P, \mathsf{T}) \to (P', \mathsf{T}')$ *is an isomorphism in* $\mathscr{D}_l^T$ *and preserves orientation.*

By localization one obtains the concept of +transversality at a point (cf. 2.1). The following important assertion then has a trivial proof (recall that preservation of orientation is a local property and that curves are connected sets).

2.13. Proposition. Let C be an oriented transversal on an oriented 2-manifold, and let $x \in C$. *Then C is a +transversal iff it is +transversal at x.*

In particular, if both $C \subset C'$ are oriented transversals, then C is a +transversal iff C' is such.

2.14. As remarked above, +transversality is independent of the particular parametrization employed in 2.11. It seems appropriate, therefore to exhibit a synthetic characterization; this is performed below, in terms of the Kronecker *intersection number* ν (Newman, 1954, VII, § 2). With later applications in view, it seemed more useful to adopt 2.11 as the definition rather than the equivalent formulation obtainable from 2.15.

Some properties of the Kronecker intersection number will be recalled. The first is that the characteristic ind familiar from the theory of functions may elegantly be defined in terms of ν (*loc. cit.*, 8.4).

2.14.1. If l is a loop in R^2 and $x \in \mathsf{R}^2$ is not on l, and if p is any path from x to the unbounded component of $\mathsf{R}^2 - \textit{image}\ l$, then

$$\operatorname{ind}_x l = \nu(p, l)\,.$$

2.14.2. In the same situation, if f is a homeomorphism of a simply connected region containing both x and l (into R^2 again), then

$$\operatorname{ind}_{f(x)} fl = \textit{degree}\,(f)\operatorname{ind}_x l$$

(*loc. cit.*, 11.1).

2.15. Proposition. Let C be an oriented curve on an oriented 2-manifold P, the carrier of a continuous ld dystem, and assume that C is transversal at x, a non-end point of C. Take $\lambda > 0$ *small enough to have*

$$x\,\mathsf{T}\,\theta \in C \quad \textit{whenever} \quad 0 < |\theta| \leq \lambda\,,$$

and let $T = x\,\mathsf{T}\,[-\lambda, \lambda]$ *be parametrized and oriented by the solution* t_x. *Then C is a +transversal iff* $\nu(T, C) = 1$.

Proof. Since the assertion is local in character, one may assume that $P = \mathsf{R}^2$. Take an orientation-preserving parametrization

$$p : [-1, 1] \approx C \quad \text{with} \quad p(0) = x\,.$$

Also consider the configuration described in Fig. 2, with p_1' the indicated loop, p' the path from $-\varepsilon i$ to εi, c' the path from $-\varepsilon$ to δ $(\varepsilon > 0 < \delta)$. Obviously

$$\nu(c', p') = \nu(c', p_1') = \operatorname{ind}_{-\varepsilon} p_1' = 1\,, \tag{5}$$

since the images of p_1' and p' intersect that of c' only within a common subarc. As the map f take the inverse to the canonic homeomorphism h corresponding to p, on taking ε, δ sufficiently small, one may assume that f is defined on the set of Fig. 2. From 2.14.2 and (5)

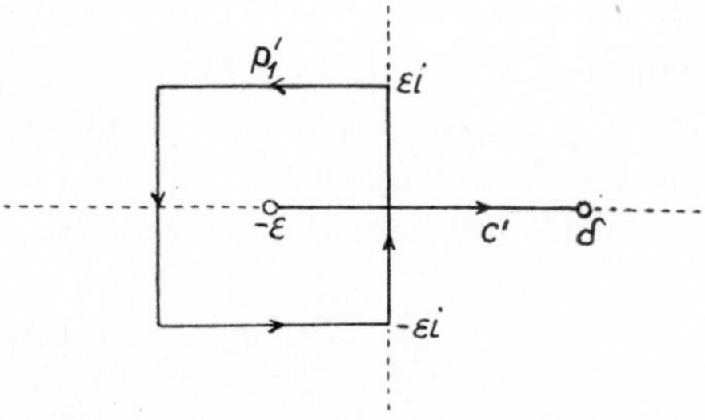

Fig. 2. Auxiliary to proof of 2.15.

$$\textit{degree}\,(f) = \operatorname{ind}_{f(-\varepsilon)} fp_1' = \nu(fc', fp_1'), \quad (6)$$

since fc' is indeed a path from $f(-\varepsilon)$ to $f(\delta)$, and since for fixed δ and small ε, $f(\delta)$ is obviously in the exterior region of the closed curve fp_1'. Now again, images of both fp_1' and fp' intersect that of fc' only in a common subarc, so that

$$\nu(fc', fp_1') = \nu(fc', fp') = \nu(T, C)\,.$$

This with (6) yields

$$\nu(T, C) = \textit{degree}\,(f)\,,$$

where, of course, $\textit{degree}\,(f) = \textit{degree}\,(f^{-1}) = \textit{degree}\,(h)$. This concludes the proof.

In 2.15, $\nu(T, C) = 1$ of course implies $\nu(T, -C) = -1$, so that T intersects C positively and $-C$ negatively. The restriction to non-end points is not essential, and again may be eliminated by section-extension as in VI,2.12. The reason for the apellation of the following theorem will be brought forward in 3.3.

2.16. Mean-value theorem. Let C be an oriented transversal of a continuous ld system on an oriented 2-manifold P. Then either all trajectories which intersect C do so positively, or all do so negatively.

Proof. 2.13, 2.15.

3. DIGRESSION: DIFFERENTIAL TRANSVERSALS

The object of this section is only to establish the connection between transversals of continuous ld systems as defined in 2.1 and the similarly named concept in classical differential equation theory; to avoid confusion, a different term will be applied to the latter.

3.1. Assume given a differential ld system T, defined by

$$\frac{\mathrm{d}z}{\mathrm{d}\theta} = f(z) \tag{1}$$

on an open subset G of R^2 (in the sense of IV,1.3.2); it is well known that the critical points of T are precisely the roots of f. Let there also be given a regular C^1 map $p : A \to G$ with A a subcontinuum of R^1; i.e. $\mathrm{d}p(\sigma)/\mathrm{d}\sigma$ is continuous and non-zero in A, so that image p is locally an arc. Assume that image p is a curve.

Then p will be said to be *differentiably transversal* to (4) at $\sigma \in A$ if

$$f(p(\sigma)) * \frac{\mathrm{d}p(\sigma)}{\mathrm{d}\sigma} \neq 0\,, \tag{2}$$

and if (2) obtains at all $\sigma \in A$, p will be termed a *differential transversal* to (1). The operator $*$ in (2) is the "exterior product" of complex numbers, defined as the real number

$$z * w = \operatorname{Im} \bar{z}w = \begin{vmatrix} \operatorname{Re} z, & \operatorname{Im} z \\ \operatorname{Re} w, & \operatorname{Im} w \end{vmatrix}$$

Obviously $(az) * (aw) = |a|^2\, z * w$, and $z * w = 0$ iff either z/w is real or $zw = 0$. Then relation (2) may then be read thus: at $p(\sigma)$, the curve *image p* does not have the same nor the opposite direction as the vector field of (1). (Cf. Hájek 1, 1960, th. 3.1 for the definition as above; and Coddington and Levinson, 1955, XVI, § 2, for rectilinear transversals.)

3.2. Proposition. With the notation of 3.1, if p is differentiably transversal to (1) at σ, then image p is transversal at p(σ) to the ld system T *associated with (1); furthermore, image p is +transversal or −transversal according to whether the expression in (2) is positive or negative.*

Proof. 3.2.1. First show that *image p* is transversal to T at $p(\sigma)$. Assume the contrary; by definition, no neighbourhood of $p(\sigma)$ in *image p* is a section of any positive t-extent, so that there then exist real θ_n, σ_n, σ_n' with

$$\sigma_n \to \sigma \leftarrow \sigma_n'\,, \quad 0 \neq \theta_n \to 0\,,$$
$$p(\sigma_n') = p(\sigma_n)\,\mathrm{T}\,\theta_n\,. \tag{3}$$

If $\sigma_n = \sigma_n'$ for infinitely many n, then $p(\sigma_n) = p(\sigma_n)\,\mathrm{T}\,\theta_n \to p(\sigma)$, and then from VI, 1.1, $p(\sigma)$ is a critical point; however, this contradicts $f(p(\sigma)) \neq 0$ in (2). Assume, therefore, that $\sigma_n \neq \sigma_n'$ for all n; from (3),

$$p(\sigma_n') - p(\sigma_n) = p(\sigma_n)\,\mathrm{T}\,\theta_n - p(\sigma_n) = \int_0^{\theta_n} \frac{\partial}{\partial\theta}(p(\sigma_n)\,\mathrm{T}\,\theta)\,\mathrm{d}\theta = \int_0^{\theta_n} f(p(\sigma_n)\,\mathrm{T}\,\theta)\,\mathrm{d}\theta\,,$$

$$\frac{p(\sigma_n') - p(\sigma_n)}{\sigma_n' - \sigma_n} = \frac{\theta_n}{\sigma_n' - \sigma_n} \cdot \frac{1}{\theta_n} \int_0^{\theta_n} f(p(\sigma_n)\,\mathrm{T}\,\theta)\,\mathrm{d}\theta\,.$$

The expression on the left has limit $\mathrm{d}p(\sigma)/\mathrm{d}\sigma$ since p is of class C^1. The second term on the right has limit $f(p(\sigma)) \neq 0$, by continuity of f, p, T and from (2). Hence $\theta_n/(\sigma_n' - \sigma_n)$ also has a finite limit, say λ, and λ is a real since all θ_n, σ_n, σ_n' are such. Thus

$$\frac{\mathrm{d}p(\sigma)}{\mathrm{d}\sigma} = \lambda f(p(\sigma))\,;$$

but then

$$f(p(\sigma)) * \frac{\mathrm{d}p(\sigma)}{\mathrm{d}\sigma} = |f(p(\sigma))|^2\, 1 * \lambda = 0\,.$$

contradicting (2). This proves 2.14.1.

3.2.2. It remains to determine whether image p (oriented by p) is a $+$*transversal* or a $-$transversal; by definition 2.6 it is required to obtain the orientation properties of the canonic homeomorphism. This latter is of the form $g_0 f_0^{-1}$ where g_0 preserves orientation and f_0 is defined by the formula

$$f_0(\theta, \sigma) = p(\sigma)\,\mathsf{T}\,\theta\,;$$

thus it suffices to compute the jacobian determinant of f_0. The jacobian of, say, $(u(x, y), v(x, y))$ is obtained readily in complex notation,

$$J(u, v) = \begin{vmatrix} \dfrac{\partial u}{\partial x}, & \dfrac{\partial y}{\partial u} \\[2ex] \dfrac{\partial v}{\partial x}, & \dfrac{\partial v}{\partial y} \end{vmatrix} = \frac{\partial}{\partial x}(u + iv) * \frac{\partial}{\partial y}(u + iv)\,.$$

In the present case then,

$$J(f_0) = \frac{\partial f_0}{\partial \theta} * \frac{\partial f_0}{\partial \sigma} \quad \text{at} \quad (0, \sigma)$$

$$= \frac{\partial}{\partial \theta}(p(\sigma)\,\mathsf{T}\,\theta)_{\theta=0} * \frac{\partial}{\partial \sigma}\,p(\sigma) = f(p(\sigma)) * \frac{\mathrm{d}p(\sigma)}{\mathrm{d}\sigma}\,.$$

Thus the sign of $J(f_0)$ coincides with that of the expression in (2); this concludes the proof of 3.2.

Conversely, of course, a transversal of the ld system associated with (1) need not be a differential transversal; however, differentiability is not the only obstacle. Indeed, consider the following

3.3. Example. Let T be the continuous gd system associated with

$$\frac{\mathrm{d}z}{\mathrm{d}\theta} = i \tag{4}$$

in R^2, and let $p(\sigma) = \sigma + i\,\varphi(\sigma)$ with continuous $\varphi : [0, 1] \to \mathsf{R}^1$. Then in any case *image* p is a section with t-extent ∞; *image* p is a regular C^1 curve if $\mathrm{d}\varphi/\mathrm{d}\sigma$ is continuous in $[0, 1]$, with infinite derivatives allowed; and p is a differential transversal of (4) if φ is of class C^1, i.e. if $\mathrm{d}\varphi/\mathrm{d}\theta$ is continuous finite.

3.4. In conclusion, the mean value theorem 2.16 will be related to the familiar mean value theorem of elementary calculus – or equivalently, to Rolle's theorem.

Assume given a map $\varphi : [0, 1] \to \mathsf{R}^1$ of class C^1 with $\varphi(0) = \varphi(1) = 0$. Define a continuous gd system in R^2 by

$$\frac{\mathrm{d}z}{\mathrm{d}\theta} = 1\,, \tag{5}$$

and the parametrization p of a regular C^1 curve by $p(\sigma) = \sigma + i\,\varphi(\sigma)$. Now postulate that p is a differential transversal, and aim at a contradiction. Then *image* p is a transversal, and thus from 2.16 all trajectories intersect *image* p positively, say. Since the trajectories of (5) are rather obvious, this implies that p passes from below each trajectory upward; in particular, *image* p intersects each trajectory at most once (a rigorous proof of this may easily be performed; for a more general case, see 4.9). But *image* p intersects the real axis at 0 and 1,

$$p(0) = 0 + i\,\varphi(0) = 0\,, \quad p(1) = 1 + i\,\varphi(1) = 1$$

which is a contradiction.

Thus p is not a differential transversal of (5), so that there exists a $\sigma \in [0, 1]$ with

$$0 = f(p(\sigma)) * \frac{\mathrm{d}p(\sigma)}{\mathrm{d}\sigma} = 1 * \left(1 + i\,\frac{\mathrm{d}\varphi(\sigma)}{\mathrm{d}\sigma}\right) = \frac{\mathrm{d}\varphi(\sigma)}{\mathrm{d}\sigma}\,.$$

This is almost Rolle's theorem, except that C^1 was assumed of φ, and only $0 \leqq \sigma \leqq 1$ was concluded.

4. NON-INTERSECTION THEOREMS

This section is devoted to a detailed study of trajectory-transversal intersections (II,6.7 also belong to this class of results). Assertions 4.4 and 4.9 describe situations in which a trajectory intersects a transversal at most once; 4.7–8 consider consequences of the presence of two intersections; 4.10 treats three intersections, and 4.11–12, 4.16 infinitely many intersections.

4.1. The carrier space P of the dynamical systems studied was assumed to be a 2-manifold; this, of course, determines the local topological structure of P completely. However, for the present results, further requirements need be imposed on the global structure of P. These are, in increasing order of strictness, the following:

that P be orientable (in the latter portions of Section 2); that P be dichotomic, a term defined below; and that P be homeomorphic to the Euclidean plane R^2 (in 4.8, 4.15).

4.2. Definition. An orientable 2-manifold P will be termed *dichotomic* if every closed curve $C \subset P$ decomposes P into two connected open sets G_1, G_2 which have C as common frontier,

$$P - C = G_1 \cup G_2\,, \quad \mathrm{Fr}\, G_1 = C = \mathrm{Fr}\, G_2\,. \quad (1)$$

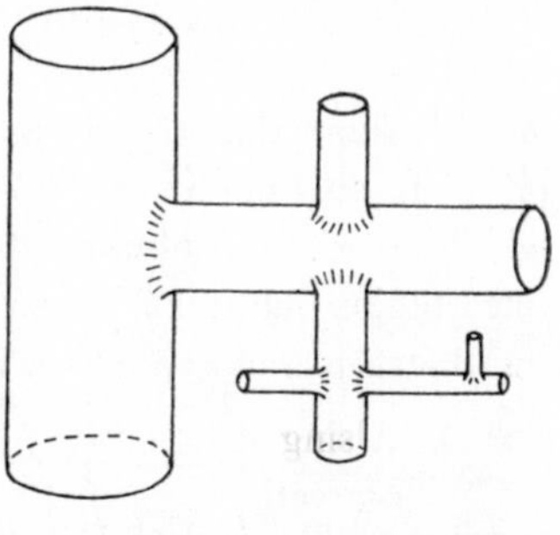

Fig. 3. Dichotomic 2-manifold.

The class of dichotomic sets includes, for example, R^2, S^2, $\mathsf{S}^1 \times \mathsf{R}^1$, but not the 2-torus $\mathsf{S}^1 \times \mathsf{S}^1$ nor the projective 2-plane; indeed, among compact 2-manifolds, only S^2 is dichotomic. Figure 3 suggests further dichotomic sets.

4.3. The results of this section may also be interpreted in another manner. Consider some property $\mathscr{V}$ proved, for example, for dichotomic carriers, and assume given a continuous ld system T on a 2-manifold P, not necessarily dichotomic. To each point of P there is an open neighbourhood $U \approx \mathsf{R}^2$, and on U the system T relativizes to a continuous ld system, which then has property $\mathscr{V}$. Thus $\mathscr{V}$ is a local property for P; it also holds on easily specifiable, and possibly large, open subsets of P, so that $\mathscr{V}$ is a relative property (in the sense of III,1.13) on dichotomic subregions of P; and if P itself is dichotomic, $\mathscr{V}$ holds on P. A single formulation then covers all three situations.

4.4. First non-intersection theorem. Let T be a trajectory and C a transversal of a continuous ld system T on a dichotomic carrier. If T or C is a closed curve, then they have one intersection point at most; if both are closed curves, they do not intersect.

The assertions of the theorem are symmetric with respect to T and C; the proof is not symmetric, and will consist of lemmas 4.5–6, where the assumptions of 4.4 are preserved (futhermore, let T and P be the ld system and carrier). Their proofs are performed in detail, since parts may serve as model in subsequent proofs.

4.5. Lemma. If C is a closed curve and intersects T, then it does so at a single point, and T is not a closed curve.

Proof. By assumption (1) holds; let U be an open neighbourhood of C which may be mapped canonically (cf. 2.4). Then C decomposes U into open sets $G_k \cap U$, and again

$$\mathrm{Fr}_U\,(G_k \cap U) = \mathrm{Fr}\, G_k \cap U = C\,.$$

Therefore, in the canonic image of U, the map of, for example, $G_1 \cap U$ is entirely inside that of C, a circle; and then the map of $G_2 \cap U$ is entirely outside this circle (since otherwise there would be common frontier points of $G_k \cap U$ near but not on C; see Fig. 1 in Section 2). Hence each $x \in C$ has the property that

$$x \mathsf{T} (-\theta) \in G_1 , \quad x \mathsf{T} \theta \in G_2 \quad \text{for small} \quad \theta > 0 . \tag{2}$$

Now assume that there exist consecutive intersections x' and $x' \mathsf{T} \theta'$ of T with C ($\theta' > 0$), so that $x' \mathsf{T} (0, \theta') \subset P - C$ is in precisely one of the G_k. Applying (2) with $x = x'$, one obtains that $x' \mathsf{T} (0, \theta')$ is in G_2; applying (2) with $x = x' \mathsf{T} \theta'$ one obtains that it is in G_1: a contradiction. A similar contradiction results on assuming that T is a closed curve, i.e. a cycle, on taking $x' = x' \mathsf{T} \theta' \in T \cap C$. This proves 4.5.

4.6. Lemma. If T is a cycle and intersects C, then it does so at a single point.

Proof. 4.6.1. Let $T = x \mathsf{T} \mathsf{R}^1$ with primitive period τ. In any case the set of intersections is finite, since each intersection is of the form

$$x \mathsf{T} \theta \in C \quad \text{with} \quad 0 \leqq \theta < \tau ,$$

and from II,6.7, the t-extent of C is a lower bound to distances of distinst θ's.

4.6.2. Using section-extension (VI,2.12 and 1.6), it may be assumed that none of the intersections is an end-point of C. This seems to be the essential trick in the present proof; the rest is a variation on the technique used in 4.5.

4.6.3. From 4.5, C is a simple arc, so that there is a parametrization

$$p : [-1, 1] \approx C ;$$

according to 4.6.1–2, notation may be so disposed that

$$p(0) = x \mathsf{T} (-\theta) , \quad p(\tfrac{1}{2}) = x \mathsf{T} \theta , \quad 0 < 2\theta < \tau ,$$

and

$$p(\sigma) \notin T \quad \text{whenever} \quad 0 < \sigma < \tfrac{1}{2} . \tag{3}$$

Now apply 2.9 with $2\theta < \varepsilon < \tau$ (otherwise preserving notation), and then 2.5. It follows that there is an open neighbourhood U of $x \mathsf{T} [-\theta, \theta]$ which may be mapped canonically; and if the closed curve T decomposes P into open sets G_1, G_2 with $\operatorname{Fr} G_k = T$, then U is decomposed into $G_k \cap U$ with

$$\operatorname{Fr}_U (G_k \cap U) = \operatorname{Fr} G_k \cap U = T .$$

Therefore, in the canonic image of U, the map of, for example, $G_1 \cap U$ is entirely below the real axis, and that of $G_2 \cap U$ is then above. Hence

$$p(\sigma) \in G_2 \quad \text{for small} \quad \sigma > 0 .$$

Then from (3), $p(\frac{1}{2} - \sigma) \in G_2$ for small $\sigma > 0$. However, since $p(\frac{1}{2}) \in T$, this contradicts 2.16. The proofs of 4.6 and 4.4 are thereby completed.

It may be observed that the proof of 4.5 also yields the following important proposition.

4.7. Proposition. On a dichotomic carrier, let $C \cup T$ be a simple closed curve with C a transversal and T a trajectorial arc. Then one component of $P - (C \cup T)$ is +invariant, the second is −invariant, and neither is invariant; hence no cycle intersects $C \cup T$.

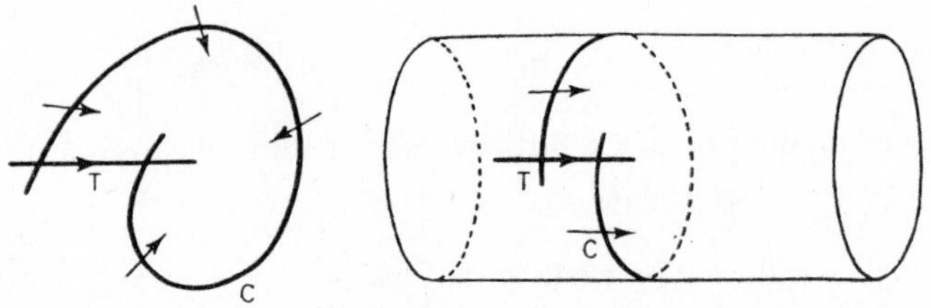

Fig. 4. Trajectory and transversal with two intersections.

Possible configurations of trajectories and transversals illustrating 4.7 are in Fig. 4. Further specialization of the carrier yields the following existence theorem.

4.8. Theorem. In a continuous ld system on R^2, the interior of each cycle, closed transversal or simple closed curve consisting of a transversal and a trajectorial arc, all contain a critical point.

Proof. For this it is necessary to anticipate the critical point existence theorem IX,4.3. The closure Q of the interior of the closed curve in question is homeomorphic to E^2, and hence is indeed triangulable with Euler characteristic 2; and it only remains to prove that Q is +invariant or −invariant. In the case of a cycle this follows from the Jordan curve theorem and IV,2.13; in the case of a closed transversal this follows from 4.5 and 2.13; and in the last case from 4.7.

4.9. Corollary. Let T be a continuous ld system on a 2-manifold P, and let $Q \subset P$ have $Q \approx \mathsf{R}^2$ and contain no critical points. Then each transversal and trajectorial arc entirely in Q intersect at one point at most.

4.10. Three-point theorem. Let C be a transversal of a continuous ld system T on a dichotomic carrier P, and assume that, for a parametrization p of C,

$$p(\sigma_k) = x \mathsf{T} \theta_k \quad (k = 1, 2, 3), \quad \theta_1 < \theta_2 < \theta_3 .$$

Then C is not a closed curve, and either $\sigma_1 = \sigma_2 = \sigma_3$ and x is on a cycle, or

$$\sigma_1 < \sigma_2 < \sigma_3 \quad or \quad \sigma_1 > \sigma_2 > \sigma_3 .$$

Proof. *4.10.1.* Applying the non-intersection theorem 4.4 repeatedly, one has the following results. C cannot be closed, since then the trajectory could not be closed, and hence would intersect C at distinct points, which is a contradiction. If $\sigma_1 = \sigma_2$ the trajectory is closed, and then the $x \mathsf{T} \theta_k$, and hence also all σ_k, coincide. Thus, consider the remaining cases: *domain* p is a subcontinuum of R^1, and the σ_k are distinct.

4.10.2. First assume that the intersections are consecutive along the trajectory. Six permutations of the σ_k's are possible; of these two are favourable, and the remaining four may be reduced to

$$\sigma_1 < \sigma_3 < \sigma_2 \quad \text{or} \quad \sigma_2 < \sigma_1 < \sigma_3$$

by changing the orientation of C; furthermore, of these the second is sent into the first by orientation-change applied to T.

4.10.3. Thus it is sufficient to exclude the case $\sigma_1 < \sigma_3 < \sigma_2$. Consider the closed curve

$$C' = (x \mathsf{T} [\theta_1, \theta_2]) \cup p[\sigma_1, \sigma_2]\,.$$

It decomposes P into regions G_1, G_2 with common frontier C'. Keeping this in mind, one sees that in the canonic image of a neighbourhood of C (cf. 2,8), the configuration must then be as indicated in Fig. 5 (this merely selects a labelling for the G_k). Obviously the trajectory through x enters G_2 at $x \mathsf{T} \theta_2$ and, since $\theta_2 < \theta_3$, subsequently exits from G_1 into G_2. Thus it must intersect the frontier C' od G_2 between $x \mathsf{T} \theta_2$ and $x \mathsf{T} \theta_3$; this contradicts the assumed consecutivity of intersections.

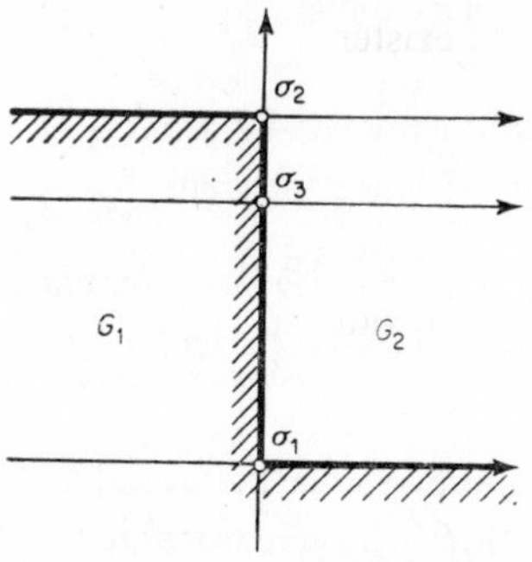

Fig. 5. Auxiliary to proof of 4.10.

4.10.4. There only remains the case where the intersections concerned are not consecutive along the trajectory. In any case, the intersections on the arc $x \mathsf{T} [\theta_1, \theta_2]$ are finite in number, according to II,6.7; the preceding proof may then be applied to all consecutive triples. This completes the proof of 4.10.

The following results are, essentially, corollaries to 4.10.

4.11. Lemma. On a dichotomic carrier, let C be a transversal, T a connected subset of a trajectory, and consider the set

$$L = \overline{C \cap T} - (C \cap T)\,.$$

Then L contains two points at most; if $C \cap T$ is infinite, then L contains a $+$limit point of T or a $-$limit point of T or both.

Proof. If either C or T is a closed curve, then $C \cap T$ is a singleton at most (cf. 4.2).

Assuming $C \cap T$ infinite, choose a parametrization $p : [0, 1] \approx C$ and an $x \in T$. From II,6.7, the set of $\theta \in \mathrm{R}^1$ with $x \mathrm{T} \theta \in C$ is of the form

$$\{\theta_n\}_n \quad \text{with} \quad \theta_n < \theta_{n+1}$$

and the n's varying over some consecutive integers. From the three-point theorem it follows that the parametrization p may be so chosen as to have

$$\sigma_n < \sigma_{n+1} \quad \text{whenever} \quad p(\sigma_n) = x \mathrm{T} \theta_n \,.$$

With this prepared, the proof proper is easily performed. If $\{\theta_n\}_n$ is infinite, it contains either $\{\theta_n\}_{n=0}^{\infty}$ or $\{\theta_{-n}\}_{n=0}^{\infty}$ or both. In the first case $\theta_n \to +\infty$, from II,6.7, whereupon the corresponding monotone sequence of σ_n's converges in $[0, 1]$, and then

$$x \mathrm{T} \theta_n = p(\sigma_n) \to p(\lim \sigma_n)$$

is a $+$limit point of x. Similarly in the second and third cases. This concludes the proof of 4.11.

The last of the intersection theorems needs a preliminary topological lemma concerning dichotomic sets, which, heuristically, is quite plausible.

4.12. Lemma. If P is a dichotomic 2-manifold and C' a closed curve in P, then each component G' of $P - C'$ is dichotomic. More precisely, if $C \subset G'$ is a second closed curve and $P - C = G_1 \cup G_2$ the decomposition into components with $C' \subset G_1$, then

$$G_2 \cup C = \bar{G}_2 \subset G' \,, \tag{4}$$

and

$$G' - C = (G_1 \cap G') \cup G_2 \tag{5}$$

is the decomposition into components with frontiers C in G'.

Proof. 4.12.1. First show that $G_2 \subset G'$. Since $\mathrm{Fr}\, G_2 = C \subset G'$ with G' open, there is $G_2 \cap G' \neq \emptyset$. If there were also $G_2 - G' \neq \emptyset$, then the connected set G_2 would also intersect $\mathrm{Fr}\, G' = C' \subset G_1$, which is a contradiction. Thus indeed $G_2 \subset G$, and hence also

$$\bar{G}_2 = G_2 \cup \mathrm{Fr}\, G_2 = G_2 \cup C \subset G'$$

as required in (4).

4.12.2. Taking intersections with G' in $P - C = G_1 \cup G_2$, one obtains (5); it is still required to prove that the summands are connected, with C as frontier in G'. Obviously

$$\mathrm{Fr}_{G'}\, G_2 = G' \cap \mathrm{Fr}\, G_2 = G' \cap C = C \,;$$

also

$$\mathrm{Fr}_{G'}(G_1 \cap G') = G' \cap \mathrm{Fr}(G_1 \cap G') \subset G' \cap (C \cup C') = C\,,$$

$$\begin{aligned}\mathrm{Fr}_{G'}(G_1 \cap G') &= G' \cap \overline{G_1 \cap G'} - (G_1 \cap G') \\ &= G' \cap \overline{\bar{G}_1 \cap G'} - G_1 \supset G' \cap \overline{C \cap G'} - G_1 = C\,.\end{aligned}$$

Thus the frontiers in G' are indeed as asserted.

4.12.3. Obviously G_2, a component of $P - C$, is connected. Finally, show that $G_1 \cap G'$ is pathwise connected. Take any x, y in $G_1 \cap G'$; then there is a path $\widehat{xy} \subset \subset G'$. If $\widehat{xy} \subset G_1$, there is nothing to prove; in the opposite case, since the end-points are in G_1, on $\widehat{xy}$ there exists a first $x_1 \in C$ and a last $y_1 \in C$. Thus there is a path $A_1 = \widehat{xx_1} \cup \widehat{x_1y_1} \cup \widehat{y_1y} \subset \bar{G}_1 \cap G$ with $\widehat{x_1y_1} \subset C$. Now, the path $\widehat{x_1y_1}$ may be contracted in the obvious manner. More explicitly, A_1 may be replaced by a path $A_2 = \widehat{xx_1} \cup \widehat{x_1y_2} \cup \widehat{y_2y}$ coinciding with A_1 except in a suitable (parametric) neighbourhood of $y_1 \in C$ within G_1, and with $\widehat{y_2y} \subset (G_1 \cap G') \cup y_2$, and $\widehat{x_1y_2}$ considerably parameter-wise shorter than $\widehat{x_1y_1}$. In a finite number of steps (corresponding to a finite open cover of $\widehat{x_1y_1}$) one obtains a path $\widehat{xx_1} \cup \widehat{x_1y} \subset (G_1 \cap G') \cup (x_1)$; and in one further step, a path $\widehat{xx_0} \cup \widehat{x_0y} \subset G_1 \cap G'$ as required. At each step the argument applied is that C decomposes a suitable neighbourhood of any $z \in C$ into two subregions of G_1 and G_2 respectively, of each of which it is the common frontier.

This concludes the proof of 4.12.

4.13. Monotonicity theorem. Assumptions: T *is a continuous ld system on a dichotomic carrier* P; T *is a +trajectory, and* C *a transversal having infinitely many intersections with* T.

Notation: $\{x_n\}_1^\infty$ *is the set of all intersections of* $C \cap T$, *in order along* T; x_∞ *is their unique limit point*; C_n *is the closed curve consisting of arcs of* C *and* T *between* x_n *and* x_{n+1}, *and* G_n *is the component of* $P - C_n$ *which contains* x_∞.

Assertion: *All* G_n *are +invariant regions with* $G_n \supset \bar{G}_{n+2}$.

Proof. Each G_n is indeed a region with C_n as frontier. From 4.7, G_n is +invariant or −invariant but not both. Now, G_n contains x_∞ and hence also x_{n+2} since $x_{n+2}\,\mathsf{T}\,\mathsf{R}^+$ does not intersect $C_n = \mathrm{Fr}\, G_n$. However, G_n does not contain x_n, which proves that G_n is not −invariant. Thus all G_n are +invariant. Finally, $G_n \supset \bar{G}_{n+2}$ follows immediately from 4.12 with $G' = G_n$, $C' = C_n$, $C = C_{n+2}$.

4.14. Remarks. At the cost of a little more effort, $G_n \supset G_{n+1}$ could be proved; however, this will not be needed later. In some cases it may be useful to select a component of $P - C_n$ in some manner other than that described in 4.13. For example, let $P \approx \mathsf{R}^2$ and let H_n be the interior of C_n (i.e. the component with compact closure). Then either

1. All H_n are −invariant regions with $\bar{H}_n \subset H_{n+2}$, or

2. There is an integer m such that 1 holds only for $n < m$, and for $n \geqq m$ all H_n are $+$invariant and $H_n \supset \overline{H}_{n+2}$. In either case all $\overline{G}_n - G_{n+2}$ are compact.

To illustrate the application of transversal theory, there is the following proof of the continuity of period theorem. This is a consequence of 4.2 (but otherwise is not connected with the topics of the present section).

4.15. Theorem. Let T *be a continuous ld system on a dichotomic carrier, let* C_n *be cycles with primitive periods* τ_n *and assume that* liminf $C_n \neq \emptyset$. *Then*

1. If $\tau_n \to 0$, *then* lim C_n *exists and is a single critical point.*

2. If liminf C_n *intersects a cycle* C_0 *with primitive period* τ_0, *then* $\tau_n \to \tau_0$ *and* lim $C_n = C_0$.

3. If liminf C_n *intersects a non-cyclic trajectory then* $\tau_n \to +\infty$.

Proof. 1 follows from VI,1.3, and 3 from VI,1.5. Consider 2; from 1 it follows that the τ_n are bounded away from zero, $0 < \delta < \tau_n$. Now take any $x \in C_0 \cap \text{liminf}\, C_n$, a transversal C through x and any $\varepsilon > 0$. Since $C \,\mathsf{T}\, [-\varepsilon, \varepsilon]$ is a neighbourhood of x (cf. 2.6), there exist $x_n \in C \cap C_n$ with $x_n \to x$. From $x \in C_0$ there follows $x \,\mathsf{T}\, \tau_0 = x$, so that also $x_n \,\mathsf{T}\, \tau_0 \to x$; and hence, in turn, $x_n \,\mathsf{T}\, \tau_0 \in C \,\mathsf{T}\, [-\varepsilon, \varepsilon]$ for large n,

$$x_n \,\mathsf{T}\, (\tau_0 + \theta_n) \in C\,, \quad |\theta_n| \leqq \varepsilon\,,$$

for some $\theta_n \in \mathsf{R}^1$. Since also $x_n \in C_n$, from 4.6 there follows $x_n = x_n \,\mathsf{T}\, (\tau_0 + \theta_n)$. As τ_n is the primitive period of $C_n \ni x_n$, there follows

$$\tau_0 + \theta_n = k_n \tau_n\,, \quad k_n \text{ integer}$$

(for large n). Now $\delta < \tau_n$, and $|\theta_n| \leqq \varepsilon$ may be made arbitrarily small; there follows $\tau_0 + \theta_n = \tau_n$, so that $|\tau_n - \tau_0| = |\theta_n| \leqq \varepsilon$, and hence $\tau_n \to \tau_0$ as required. This completes the proof of 4.15.

A similar proof (but without recourse to 2.6) may be used for the following result, which yields more information on the situation described in the monotonicity theorem.

4.16. Lemma. Let $x \,\mathsf{T}\, \theta_n$ *be consecutive intersections with a transversal* C *(on a dichotomic carrier), and* $x \,\mathsf{T}\, \theta_n \to x_\infty$ *with* $\theta_n \to +\infty$. *Then either*

1. x_∞ *is on a cycle with primitive period* τ *and* $\theta_{n+1} - \theta_n \to \tau$, *or*

2. x_∞ *is on a non-cyclic trajectory, and* $\theta_{n+1} - \theta_n \to +\infty$.

4.17. Examples.

4.17.1. In 4.15.2 it is essential that the dimension of the carrier manifold be 2. Thus, consider the fourth-order linear differential equation

$$y^{\mathrm{IV}} + 5y^{\mathrm{II}} + y = 0\,;$$

its 2π-periodic solutions $\varepsilon \sin \theta + \sin 2\theta$ converge for $\varepsilon \to 0$ to a π-periodic solution $\sin 2\theta$. The carrier of the corresponding first-order autonomous equation is R^4, with dimension 4. Slightly more effort yields an example in R^3 exhibiting similar phenomena.

4.17.2. If, in 4.15, liminf C_n contains critical points, then neither $\tau_n \to \tau_0 \neq 0$ nor $\tau_n \to +\infty$ can be excluded. For example, the gd systems in R^2 associated with

$$\frac{dz}{d\theta} = iz \quad \text{and} \quad \frac{dz}{d\theta} = i|z|\, z$$

have cycles $|z| = \varrho$ for arbitrary $\varrho > 0$, with periods 2π and $2\pi/\varrho$ respectively.

4.17.3. Take any positive irrational α, and consider the continuous gd system defined in R^2 by

$$\frac{dz}{d\theta} = 1 + i\alpha\,;$$

also the equivalence relation $z_1 \sim z_2$ if $z_1 - z_2$ has integral real and imaginary parts. Then $R^2/\sim$ is the familiar model of the 2-torus $S^1 \times S^1$, and since the assumptions of IV,3.8 are satisfied, there results a continuous gd system on $S^1 \times S^1$. This latter is easily described: if S^1 is identified with $[0, 1)$ and $\{\theta\}$ denotes the fractional part of θ, then the solution through $a = x + iy$ is

$$t(\theta) = \{x + \theta\} + i\{y + \alpha\theta\}\,, \quad \theta \in R^1\,. \tag{6}$$

To prove that each trajectory is dense in $S^1 \times S^1$, it is sufficient to show that the complex numbers (6) (with x, y, α fixed and θ variable) are dense in $[0, 1] \times [0, 1]$; or, by setting $\theta = -x + n$ with n integral, that

$$\{y - \alpha x + \alpha n\}\,, \quad n \in C^1\,.$$

are dense in $[0, 1]$. Since α is irrational, this is a well known result (for example, Franklin, 1947, I, exer. 32, for the case $y - \alpha x = 0$).

CHAPTER VIII

Limit Sets on 2-manifolds

The main and familiar application of transversals is in the study of limit sets; the possible application to perturbation theory is not performed in the present book.

Section 1 presents the classical results on the dynamical and topological constitution of limit sets. In particular, in 1.9, 1.11, 1.13–14 there may be recognized the Poincaré-Bendixson theory (as in, for example, Coddington and Levinson, 1955, XVI), carried over to continuous ld systems on 2-manifolds – with minor generalisations (for example, in 1.13 absence of only the +asymptotic and not all critical points is required). In the same sense, 1.5 may be compared with Coddington and Levinson (1955, XVI, remark to 2.3); 1.18 with a theorem due to Vinograd (Niemyckij and Stepanov, 1949, II, § 1, th. 8); and 1.21 with a result of Sol'ncev (cf. Niemyckij, 1954, p. 425 with obviously incomplete assumptions).

Section 2 concerns the orbital stability properties of limit sets. The present treatment of this and the preceding sections follows that of Hájek 8 (1965, §§ 5–6). In Sections 3 and 4 stronger results are obtained on restricting the limit sets studied. This includes the analogue to Poincaré's celebrated theorem on consecutive cycles 3.5 (Lefschetz, 1957, X, § 5), and the characterization of ao+ stable points in 4.6.

Proposition 4.9 excepted, isolated critical points with no stability properties assumed are not treated. Recently, Nagy has developed index methods powerful enough to apply to continuous ld systems on manifolds and to yield the theorems familiar from differential equation theory (see Nagy, 1965).

The convention of Chapter VII, that there is given a continuous ld system T on a 2-manifold P, is also reproduced for the purposes of the present chapter.

1. STRUCTURE OF LIMIT SETS

The definition and elementary properties of limit sets were presented in IV,4. On restricting the carrier spaces to dichotomic 2-manifolds, considerably deeper results may be obtained. The basic lemmas of this section are 1.2–3; there follow properties of general +limit sets including the Poisson stability theorem, and of compact +limit sets, for example the Poincaré-Bendixson theorem. In conclusion, the topological structure of non-critical +limit sets is discussed.

1.1. For convenience of formulation, the following terminology will be introduced. A critical point $x \in P$ is termed *+asymptotic* if $x \in L_y^+$ for some $y \neq x$; a cycle $T \subset P$ is called a *+limit cycle* if $T \subset L_y^+$ for some $y \notin T$. By orientation change one obtains the concepts of −asymptotic points and −limit cycles; a *limit cycle* is to mean a +limit or −limit cycle, an *asymptotic point* is a +asymptotic or −asymptotic point. A +limit set (or −limit or limit set) is termed *critical* if it consists exclusively of critical points, for example, if it is void), and non-critical otherwise.

1.2. Lemma. In a dichotomic carrier let $y \top \theta_n \to x$ with $\theta_n \to +\infty$ and x non-critical, and let C be any transversal containing x as non-end-point. Then there exist $\theta_n' \in \mathsf{R}^1$ with

$$y \top \theta_n' \in C, \quad \theta_n' - \theta_n \to 0 \tag{1}$$

and

$$y \top \theta_n' \to x, \quad \theta_n' \to +\infty .$$

Proof. According to VII,2.5, $C \top (-\varepsilon, \varepsilon)$ is a neighbourhood of x for any $\varepsilon > 0$. Since $y \top \theta_n \to x$, one concludes that $y \top \theta_n \in C \top (-\varepsilon, \varepsilon)$ for large n, so that there must exist uniquely determined $\theta_n' \in \mathsf{R}^1$ with $y \top \theta_n' \in C$ and $|\theta_n' - \theta_n| < \varepsilon$ (the first few θ_n' may be taken arbitrarily). As ε may be taken arbitrarily small, there follows $\theta_n' - \theta_n \to 0$; this proves (1). Hence

$$y \top \theta_n' = (y \top \theta_n) \top (\theta_n' - \theta_n) \to x \top 0 = x$$

and

$$\theta_n' = \theta_n + (\theta_n' - \theta_n) \to +\infty ,$$

concluding the proof of 1.2.

1.3. Lemma. In a dichotomic carrier P, assume that $x \in L_a^+ \cap C$ for some a, $x \in P$ and transversal $C \subset P$. Then, for all $\theta \in \mathsf{R}^1$,

$$L_a^+ \cap (C \top \theta) = x \top (\theta) . \tag{2}$$

Proof. One inclusion in (2) is obvious: from II,2.13,

$$x \top (\theta) \subset (L_a^+ \cap C) \top \theta \subset (L_a^+ \top \theta) \cap (C \top \theta) = L_a^+ \cap (C \top \theta)$$

since L_a^+ is invariant (cf. IV,4.6) and hence $L_a^+ \top \theta = L_a^+$. Next, take any $x' \top \theta \in L_a^+$ with $x' \in C$; then $a \top \theta_n \to x' \top \theta$ for some $\theta_n \to +\infty$, and hence

$$a \top (\theta_n - \theta) \to (x' \top \theta) \top (-\theta) = x' \in L_a^+ .$$

Thus x and x' are both in $L_a^+ \cap C$, i.e. both are +limit points of a on C; then VII,4.11 in conjunction with 1.2 yields $x = x'$, and hence $x \top \theta = x' \top \theta$. This verifies the second inclusion needed for (2), and thus completes the proof of 1.3.

1.4. Lemma 1.2 makes it possible to eliminate express mention of any specific transversal in several of the assertions of VII,4; thus for example the conclusion of VII,4.16.2 obtains if it is assumed that, on a dichotomic carrier, $x \mathsf{T} \theta_n \to x_\infty$ with $\theta_n \to +\infty$ and x_∞ non-critical. VII,4.11 was used in a similar manner to obtain 1.3. We proceed to exploit these two lemmas rather intensively.

1.5. Second non-intersection theorem. Let C be a transversal of a continuous ld system on a dichotomic carrier. Then C intersects each +limit set in one point at most; if C is closed, it intersects no +limit set.

Proof. The first assertion is 1.3 with $\theta = 0$ (recall that all points on a transversal are non-critical). For the second apply 1.2 to obtain an infinite set of intersections with a trajectory, contradicting the first intersection theorem VII,4.4.

1.6. Theorem. For a continuous ld system on a dichotomic carrier, the only Poisson +stable sets are the cycles, the critical points, and the empty set.

Proof. According to the definition, IV,4.8, it is required to prove the following: if

$$a \mathsf{T} \theta_n \to a, \quad \theta_n \to +\infty,$$

then a is critical or on a cycle.

Assume the contrary; in particular, then, a is non-critical, and VII,2.6 may be applied, yielding a transversal C with parametrization $p : [-1, 1] \approx C$ and $p(0) = a$. Now apply 1.2, obtaining a sequence θ'_n as indicated there. Since by assumption $a \mathsf{T} \mathsf{R}^1$ is not a cycle, to each θ'_n there is precisely one σ_n with

$$p(\sigma_n) = a \mathsf{T} \theta'_n, \quad |\sigma_n| \leqq 1;$$

and in particular $p(0) = a = a \mathsf{T} 0$. From the three-point theorem VII,4.10 it then follows that

$$\theta'_m < \theta'_n \quad \text{implies} \quad \sigma_m < \sigma_n$$

(on orienting C appropriately). As $\theta'_n \to +\infty$, one has $0 < \theta'_m < \theta'_n$ for large $n > m$; and then in turn

$$0 < \sigma_m < \sigma_n, \quad 0 < \sigma_m < \lim \sigma_n,$$

$$a = p(0) \neq p(\lim \sigma_n) = \lim p(\sigma_n) = \lim a \mathsf{T} \theta'_n = a.$$

This contradiction concludes the proof of 1.6.

The preceding result treated trajectories intersecting their +limit set; the following considers trajectories whose +limit and −limit sets intersect.

1.7. Lemma. On a dichotomic carrier, assume that $L_a^+ \cap L_a^-$ contains a non-critical point; then a is on a cycle and $L_a^+ = a \mathsf{T} \mathsf{R}^1 = L_a^-$.

Proof. 1.2 applied to VII,4.11.

1.8. Lemma. On a dichotomic carrier, L_a^+ is an arc locally at each of its non-critical points; if x_i are on distinct trajectories in L_a^+ and $x_i \to x$, then x is a critical point.

Proof. Let x' be a non-critical point in L_a^+; applying VII,2.6 at x' and then 1.3, (2) yields an arc

$$L_a^+ \cap (C \top [-\varepsilon, \varepsilon]) = x' \top [-\varepsilon, \varepsilon] \tag{3}$$

with $C \top [-\varepsilon, \varepsilon]$ a neighbourhood of x', i.e. the first assertion. The second also follows from (3), concluding the proof of 1.7.

The conclusions in 1.8 are local in character. Since each point on a 2-manifold has dichotomic neighbourhoods (indeed, homeomorphs of R^2), it might be thought that the dichotomicity assumption could be omitted. Example VII,4.17.3 shows that this is not the case: $L_a^+ = \mathsf{S}^1 \times \mathsf{S}^1$ is not locally an arc.

1.9. Theorem. For a continuous ld system on a dichotomic carrier P, every +limit set consists of critical points and of a countable system of non-critical trajectories T_n; each compact subset of P without critical points intersects only a finite number of the T_n.

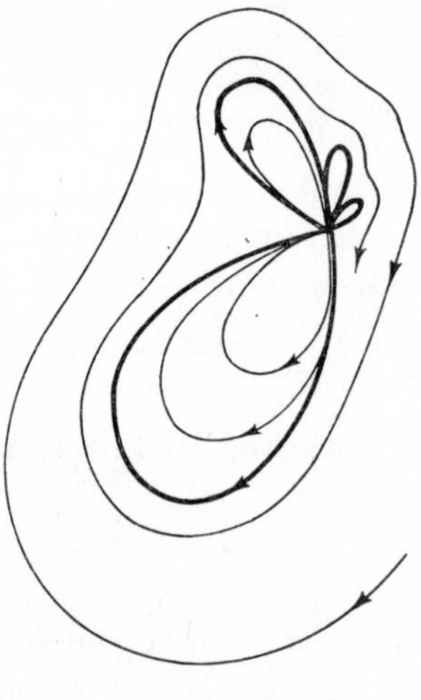

Fig. 1. +Limit set containing infinitely many trajectories.

Proof. From 1.8, each non-critical trajectory in L_a^+ is open in L_a^+ and, of course, distinct trajectories are disjoint. Since L_a^+ has a countable dense subset (cf. IV,4.4), the system of such trajectories is countable. The last assertion again follows from 1.8.

1.10. Example. In 1.9, the number of non-critical trajectories in L_a^+ may well be infinite. An analytic example would require some effort, but possibly it will suffice to indicate the principle of construction in Fig. 1.

1.11. Proposition. For each +limit set in a dichotomic carrier the following alternative obtains: either L_a^+ is a cycle or both L_x^+ and L_x^- are critical for all $x \in L_a^+$.

Proof. 1.11.1. First show that any cycle within L_a^+ necessarily coincides with L_a^+. In proving this it suffices to assume that $a \top \mathsf{R}^1$ itself is not a cycle; let L_a^+ contain a cycle $x \top \mathsf{R}^1$ with primitive period τ, and take a transversal C through x as in VII,2.6. Since $x \in x \top \mathsf{R}^1 \subset L_a^+$, from 1.2 and non-cyclicity of $a \top \mathsf{R}^1$ it follows that there are infinitely many intersections of $a \top \mathsf{R}^+$ with C; let these be $a \top \theta_n$ with $a \top \theta_n \to x$ and with $\theta_n \to +\infty$ monotonously as $n \to +\infty$. According to VII,4.16,

$$\theta_{n+1} - \theta_n \to \tau . \tag{4}$$

To prove $L_a^+ \subset x \top \mathsf{R}^1$ take any $x' \in L_a^+$, so that one has

$$a \top \theta'_n \to x' \quad \text{with} \quad \theta'_n \to +\infty\,;$$

then to each (sufficiently large) n there is a k_n with

$$\theta_{k_n} \leqq \theta'_n < \theta_{k_n+1}\,, \quad k_n \to +\infty\,.$$

From (4), the sequence $\theta'_n - \theta_{k_n}$ is bounded, so that one may take a convergent subsequence with indices $m = m_n \to +\infty$,

$$\theta'_m - \theta_{k_m} \to \lambda \quad (0 \leqq \lambda \leqq \tau)\,.$$

Then

$$x' = \lim a \top \theta'_n = \lim a \top \theta'_m = \lim \left((a \top \theta_{k_n}) \top (\theta'_m - \theta_{k_m})\right) = x \top \lambda$$

is indeed in $x \top \mathsf{R}^1$ as required.

1.11.2. To prove 1.11, assume the contrary. Then there exist a, x, x' such that L_a^+ is non-cyclic, $x \in L_a^+$, $x' \in L_x^+ \cup L_x^-$ and x' is non-critical. According to 1.11.1, $x \top \mathsf{R}^1$ is non-cyclic, since otherwise L_a^+ would be a cycle. Now construct a transversal C to x' as in VII,2.6; from $x' \in L_x^+ \cup L_x^+$ it follows that $x \top \mathsf{R}^1$ intersects C infinitely often, so that, using $x \top \mathsf{R}^1 \subset L_a^+$, L_a^+ also intersects C infinitely often. However, this contradicts the second non-intersection theorem, and concludes the proof of 1.11.

1.12. Using the terminology of 1.1 and theorem 1.6, the preceding result may also be formulated thus: For every point a in a dichotomic carrier there obtains precisely one of the following cases: a is critical, or $a \top \mathsf{R}^1$ is a cycle, or L_a^+ is a +limit cycle, or both L_x^+ and L_x^- consist of +asymptotic points for all $x \in L_a^+$.

The following results concern +limit sets of +trajectories with compact closure (i.e. of Lagrange +stable semi-trajectories). In this case IV,4.7 and VI,3.12–14 are available; and there follow the two classical theorems below.

1.13. Poincaré theorem. Given a continuous ld system on a dichotomic carrier P, and a closed subset $Q \subset P$; if Q contains a non-critical Lagrange +stable +trajectory but no +asymptotic points, then Q contains a cycle; namely, the +limit set of the +trajectory.

1.14. Poincaré-Bendixson theorem. Let $\top$ be a continuous ld system on a dichotomic carrier P, and take any Lagrange +stable non-critical $a \in P$. Then either $a \top \mathsf{R}^1$ is a cycle or L_a^+ is a +limit cycle or, for every $x \in L_a^+$, both L_x^+ and L_x^- are non-void compact connected sets of +asymptotic points.

Proofs. IV,4.7, 1.12.

1.15. Corollary. If in 1.14 the set of +asymptotic points in P is isolated, then the last alternative may be altered to read: for each $x \in L_a^+$, *both* L_x^+ *and* L_x^- *are single +asymptotic points.*

1.16. The concluding group of results concerns the topological properties of compact non-critical +limit sets. That there is some analogy between these and closed curves has already been suggested by similarities in the first and second non-intersection theorems. In 1.18 and 1.21 another parallel is established, in connection with the property of closed curves in R^2 described in the Jordan curve theorem.

1.17. Lemma. On a dichotomic carrier, each +limit set is nowhere dense.

Proof. Assume the contrary. Since +limit sets are closed, there exists an L_x^+ and a non-void open $G \subset L_x^+$. Then $x \mathsf{T} \theta \in G$ for some $\theta \in \mathsf{R}^+$, and hence

$$L_x^+ \supset (x \mathsf{T} \mathsf{R}^+) \cap L_x^+ \supset (x \mathsf{T} \mathsf{R}^+) \cap G \ni x \mathsf{T} \theta ;$$

from 1.6 then, $x \mathsf{T} \mathsf{R}^+$ is a cycle or critical and therefore $L_x^+ = x \mathsf{T} \mathsf{R}^+$ is nowhere dense. This contradiction proves 1.17.

1.18. Proposition. In a dichotomic carrier P, let $a \mathsf{T} \mathsf{R}^1$ *be non-cyclic Lagrange +stable. If*

$$P - L_a^+ = D_a \cup \bigcup_{i \neq a} D_i \tag{5}$$

is the decomposition into components with $a \in D_a$, *then* D_a *is open, simply connected and* $\operatorname{Fr} D_a = L_a^+$.

Remarks. Each D_j is connected by definition. Since $P - L_a^+$ is open, and P a 2-manifold and hence locally connected, each D_j is open (Kuratowski, 1950, § 44, II,4). As L_a^+ is invariant, each D_j is also such (cf. IV,4.6, II,5.6.5, IV,2.13). Nothing is asserted concerning the number of non-void D_j's. From 1.6 and the assumption on $a \mathsf{T} \mathsf{R}^1$ it follows that a is indeed in some component D_j, and $a \in D_a$ then merely fixes the notation; from invariance it then follows, of course, that $a \mathsf{T} \mathsf{R}^1 \subset D_a$.

Proof of 1.18. Since P is locally connected and D_j is a component of $P - L_a^+$, from Kuratowski (1950, § 44, III, 3) it follows that

$$\operatorname{Fr} D_j \subset \operatorname{Fr} (P - L_a^+) ;$$

hence and from 1.17,

$$\operatorname{Fr} D_j \subset \operatorname{Fr} (P - L_a^+) = \operatorname{Fr} (L_a^+) = L_a^+ . \tag{6}$$

Since $L_a^+ \subset \overline{a \mathsf{T} \mathsf{R}^+} \subset \bar{D}_a$ and $L_a^+ \cap D_a = \emptyset$, one also has

$$L_a^+ \subset \bar{D}_a - D_a = \operatorname{Fr} D_a ;$$

this and (6) yield $\mathrm{Fr}\, D_a = L_a^+$ as asserted. It remains to show that $P - D_a$ is connected; to this end consider

$$P - D_a = L_a^+ \cup \bigcup_{i \neq a} D_i$$

obtained from (5). All summands on the right are connected (for L_a^+ see IV,4.7). Since P is connected and $D_i \subset P - L_a^+ \neq P$, there is $\mathrm{Fr}\, D_i \neq \emptyset$ for all $D_i \neq \emptyset$, so that from (6)

$$\bar{D}_i \cap L_a^+ \supset \mathrm{Fr}\, D_i \cap L_a^+ = \mathrm{Fr}\, D_i \neq \emptyset ;$$

this shows that D_i and L_a^+ are not separated. Using the theorem on unions of connected sets (Kuratowski, 1950, § 41, II, 2), it follows that $P - D_a$ is indeed connected. This concludes the proof of 1.17.

1.19. To obtain further results on the situation studied in 1.18, additional apparatus is required, namely the concepts introduced in the monotonicity theorem.

Assume again that $a \mathsf{T} \mathrm{R}^1$ is non-cyclic and Lagrange +stable in a dichotomic carrier P; and that there are given a non-critical point $x \in L_a^+$ and a transversal C containing x as non-end-point (if x exists then C exists: VII,2.6). From 1.2, $a \mathsf{T} \mathrm{R}^+$ intersects C infinitely often; the monotonicity theorem VII,4.13 may then be applied, yielding regions G_n and closed curves $C_n = \mathrm{Fr}\, G_n$ with the properties described there.

1.20. Lemma. With the assumptions and notation of 1.19 and 1.18,

$$D_a = \bigcup (P - \overline{G_{2n}}) = \bigcup (P - G_{2n}) , \tag{7}$$

$$L_a^+ = \lim C_{2n} . \tag{8}$$

Proof. 1.20.1. The displayed relations will be proved in several intermixed steps. First,

$$\operatorname{limsup} C_{2n} \subset \overline{\bigcap_{m \geq 1} \bigcup_{n \geq m} C_{2n}} \subset L_a^+ \tag{9}$$

from the construction of the C_n in IV,4.13.

1.20.2. Now consider (7). Take any G_{2n}; since $a \mathsf{T}\, \theta \in G_{2n}$ for some $\theta \in \mathrm{R}^+$ and G_{2n} is +invariant, there is

$$L_a^+ \subset \overline{a \mathsf{T} [\theta, +\infty)} \subset \overline{G_{2n}} ;$$

hence and from (5),

$$P - \overline{G_{2n}} \subset P - L_a^+ = D_a \cup \bigcup_{i \neq a} D_i$$

with $P - \overline{G_{2n}}$ connected. Thus $P - \overline{G_{2n}}$ is in some unique D_j; and since $a \mathsf{T} \mathrm{R}^1 \subset D_a$

and $P - \overline{G_{2n}}$ contains a subarc of $a \mathsf{T} \mathsf{R}^1$, this implies that $P - \overline{G_{2n}} \subset D_a$. This proves

$$\bigcup (P - \overline{G_{2n}}) \subset D_a \,. \tag{10}$$

1.20.3. The proof of the converse inclusion is slightly more involved. Take any $y \in D_a$, postulate $y \in \bigcap \overline{G_{2n}}$ and aim at a contradiction. Since both a and y are in D_a, there is an arc $\widehat{ay} \subset D_a$. From the construction of the C_n one has $a \notin \overline{G_{2n}}$, and $y \in \overline{G}_{2n}$ by assumption; thus there exists a $y_n \in \operatorname{Fr} G_{2n} = C_{2n}$ on $\widehat{ay}$. Since the latter set is compact, it may even be assumed that $y_m \to y_0 \in \widehat{ay} \subset D_a$ for some subsequence; and from $y_m \in C_{2n}$ and (9) there then follows $y_0 \in L_a^+$. Thus y_0 is in both D_a and L_a^+, contradicting (5). This proves the inclusion converse to (10); since also $G_n \supset \overline{G_{n+2}}$, one concludes (7).

1.20.4. Since always $\operatorname{Fr} \bigcup X_i \subset \overline{\bigcup \operatorname{Fr} X_i}$ in a locally connected space (Kuratowski, 1950, § 44, III, 1), from (7) there follows

$$\operatorname{Fr} D_a = \operatorname{Fr} \bigcup (P - \overline{G_{2n}}) \subset \overline{\bigcup \operatorname{Fr} (P - \overline{G_{2n}})} \,;$$

or, on applying 1.18 and $C_n = \operatorname{Fr} (P - \overline{G_n})$,

$$L_a^+ \subset \overline{\bigcup C_{2n}} \,.$$

More generally, from (7) and monotonicity of the G_n,

$$L_a^+ \subset \overline{\bigcup_{n \geqq m} C_{2n}}$$

for each $m > 1$. This implies $L_a^+ \subset \liminf C_{2n}$, and with (9) concludes the proof of (8) and of 1.20.

1.21. Theorem. Each compact non-critical +limit set of a continuous ld system on a dichotomic carrier separates the carrier.

Proof. Let L_a^+ be a +limit set as described; in particular, $a \mathsf{T} \mathsf{R}^+$ is non-critical and Lagrange +stable according to VI,3.13. The conclusion manifestly holds if $a \mathsf{T} \mathsf{R}^+$ is a cycle; therefore assume $a \mathsf{T} \mathsf{R}^+$ is non-cyclic, so that the premises of 1.18–19 are satisfied. Now assume that the conclusion does not hold; then (5) reduces to $P - L_a^+ = D_a$, and from (7)

$$L_a^+ = \bigcap G_{2n} \,,$$

on choosing a non-critical $x \in L_a^+$ and then constructing C, G_n, C_n as described in 1.18. In a canonic neighbourhood of x, the frontier C_{2n} of G_{2n} would then have to intersect C in a manner inconsistent with the three-point theorem (VII,4.3; viz on both sides of x). This contradiction concludes the proof of 1.21.

Remark. The non-criticality assumption in 1.21 is essential: otherwise L_a^+ might well be a single critical point or a critical arc, which then, of course, does not separate the carrier space. For a similar reason in 1.20 it cannot be concluded that

$$\overline{\bigcup_{i \neq a} D_i} = \bigcap \overline{G_{2n}},$$

since from (7) and (5)

$$\bigcap \overline{G_{2n}} = P - D_a = L_a^+ \cup \bigcup_{i \neq a} D_i,$$

and one may easily have

$$\overline{\bigcup D_i} \subsetneq L_a^+ \cup \bigcup D_i$$

(see Fig. 2).

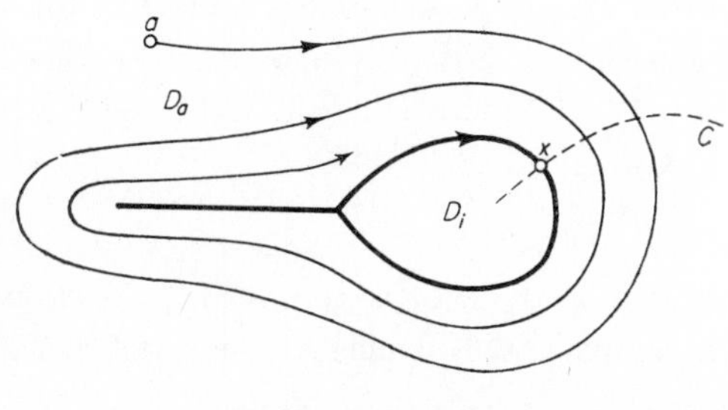

Fig. 2. Non-critical +limit set with critical arc.

2. STABILITY OF LIMIT SETS

If a +trajectory T spirals towards a +limit cycle T' as its +limit set, it seems reasonable to expect that similar behaviour will be exhibited by all +trajectories near T' and in the same component D of $P - T'$ as T; or, more precisely, that T' is asymptotically orbitally +stable relative to $\bar{D}$ (cf. IV,4.9; with D invariant, the term "relative" concerns the relativization of the original ld system in the sense of III,1.13 and IV,2.9). This is indeed the case, and the property will be demonstrated for a considerably more general class of +limit sets.

2.1. Throughout this section the following situation is studied. There are given, in addition to the continuous ld system T on a dichotomic carrier P, a Lagrange +stable point $a \in P$ such that $a \mathsf{T} \mathsf{R}^1$ is not a cycle and that there exists a non-critical point $b \in L_a^+$. A transversal C to b is then taken as in VII,2.6, and the construction of the monotonicity theorem performed. This brings into action the apparatus of 1.18–19, the regions D_j and G_n and the closed curves C_n; this notation will be preserved. Also define

$$V_n = \overline{G_{2n}} \cap \overline{D_a}.$$

2.2. Lemma. In the situation of 2.1, the V_n form a decreasing sequence of closed +invariant sets, compact for n sufficiently large; furthermore,

$$\overline{G_{2n}} \subset V_n \cup (P - D_a), \tag{1}$$

$$L_a^+ = \bigcap V_n. \tag{2}$$

Proof. Obviously

$$V_n \cup (P - D_a) = (V_n \cap D_a) \cup (P - D_a) = (\overline{G_{2n}} \cap D_a) \cup (P - D_a) \supset$$
$$\supset (\overline{G_{2n}} \cap D_a) \cup (\overline{G_{2n}} - D_a) = \overline{G_{2n}},$$

i.e. (1). For (2) apply 1.20 and then 1.18:

$$\bigcap V_n = \overline{D_a} \cap \bigcap \overline{G_{2n}} = \overline{D_a} \cap (P - \bigcup(P - \overline{G_{2n}})) = \overline{D_a} \cap (P - D_a) = \operatorname{Fr} D_a = L_a^+ .$$

From the properties of G_{2n} and D_a it follows that the V_n indeed form a decreasing sequence of closed +invariant sets. Since L_a^+ is compact non-void (assumptions and IV,4.7) there is a compact neighbourhood U of L_a^+. From (2)

$$L_a^+ = \lim V_n ,$$

so that $V_n \subset U$ for large n; thus V_n is closed in a compact set U, and hence V_n itself is compact. This concludes the proof of 2.2.

2.3. Theorem. In a continuous ld system T *on a dichotomic carrier* P*, let* $a \mathsf{T} \mathsf{R}^+$ *be non-cyclic with* L_a^+ *compact non-critical. Then* L_a^+ *is o+ stable relative to the closure of that component of* $P - L_a^+$ *which contains* a*.*

Proof. From VI,3.11, a is Lagrange +stable, so that 2.1–2 apply; then

$$L_a^+ \subset \overline{G_{2n+2}} \subset G_{2n} ,$$

so that the $V_n = \overline{G_{2n}} \cap \overline{D_a}$ are arbitrarily small +invariant neighbourhoods of L_a^+ in $\overline{D_a}$. The assertion then follows directly from the definition IV,4.9.

2.4. Example. In 2.3, L_a^+ may well not be *ao+* stable relative to $\overline{D_a}$. To see this, first take the differential equation

$$\frac{d\varrho}{d\varphi} = \varrho(\varrho - 1)^2$$

with ϱ, φ considered as polar coordinates in R^2; more precisely, consider the differential ld system in R^2 defined by

$$\frac{dz}{d\theta} = z(i + (|z| - 1)^2) . \tag{3}$$

There is a unique critical point, at 0, and a unique cycle, the unit circle K; all other trajectories spiral towards or away from 0 or K or ∞. Now modify this system by using

$$\frac{dz}{d\theta} = z(i + (|z| - 1)^2)\, \psi(z)$$

with

$$\psi(x + iy) = (x - 1)^2 \sin^2 \frac{1}{x - 1} + y^2$$

($x, y \in \mathsf{R}^1$), so that ψ is continuous non-negative. The modified system has further critical points at the zeros of ψ,

$$\left(1 \pm \frac{1}{\pi}, 0\right), \quad \left(1 \pm \frac{1}{2\pi}, 0\right), \ldots$$

but all other trajectories are unchanged (compare with VI,1 11). In particular, some trajectory still spirals towards K as its compact non-critical +limit set (K is no longer a cycle). Thus the assumptions in 2.3 are satisfied, but K is not $ao+$ stable, since there are critical points arbitrarily near K in each component of $\mathsf{R}^2 - K$.

The second principal result of this section is theorem 2.9 concerning asymptotic orbital stability of limit sets; 2.5–7 are intermediate partial results. It will be shown that example 2.4 is typical of non-asymptotic orbital +stability in that the presence of critical points is the sole obstacle.

2.5. Lemma. In the situation of 2.1–2 if V_1 is compact and there are no +asymptotic points nor cycles in $V_1 - L_a^+$, then V_1 is a region of attraction for L_a^+.

Proof. For each $x \in V_1$ there are, principally, two possibilities. Either $x \mathsf{T} \mathsf{R}^+$ intersects each V_n, whereupon

$$L_x^+ \subset \overline{x \mathsf{T} \mathsf{R}^+} \subset \overline{V_n} = V_n$$

from +invariance, and

$$L_x^+ \subset \bigcap V_n = L_a^+ \tag{4}$$

according to 2.2. Or there is $x \mathsf{T} \mathsf{R}^+ \subset P - V_n$ for some n, whereupon

$$x \mathsf{T} \mathsf{R}^+ \subset V_1 - V_n = (\overline{G_2} \cap \overline{D_a}) - (\overline{G_{2n}} \cap \overline{D_a}) = \overline{D_a} \cap (\overline{G_2} - \overline{G_{2n}}) \subset V_1 - G_{2n},$$

and $V_1 - G_{2n}$ is a compact subset of $V_1 - L_a^+$. With our assumptions on $V_1 - L_a^+$, this contradicts the Poincaré theorem 1.13. Thus (4) holds for all $x \in V_1$ as asserted.

2.6. Lemma. If, under the assumptions of 2.1, L_a^+ is a cycle (in other words, if L_a^+ is a +limit cycle of a on a dichotomic carrier), then some V_n is a region of attraction for L_a^+.

Proof. The set of critical points is closed (VI,1.2), and each cycle is a compact set containing no critical points; therefore there is a neighbourhood of L_a^+ without critical points. There are no cycles in D_a near L_a^+, since for example from VII,2.9) such cycles would intersect C and hence some C_n, in contradiction with VII,4.7. Thus from 2.2, for sufficiently large n, V_n is compact and there are no cycles nor critical points in

$$V_n \cap D_a = \overline{G_{2n}} \cap D_a = \overline{G_{2n}} \cap (\overline{D_a} - \mathrm{Fr}\, D_a) = \overline{G_{2n}} \cap \overline{D_a} - L_a^+ = V_n - L_a^+ ;$$

the assertion then follows on applying 2.5.

2.7. *Proposition. If T is a cycle on a dichotomic carier, then the set of points $x \in P - T$ with $L_x^+ = T$ is open.*

Proof. Take any $a \notin T$ with $L_a^+ = T$; then $a \mathsf{T} \mathsf{R}^+$ is noncyclic Lagrange +stable (cf. 1.6 and VII,3.11) and of course L_a^+ is non-critical. Construct D_a and the V_n as described in 2.1, and take some V_n as indicated in 2.6. Since V_n is a neighbourhood of L_a^+ in $\overline{D_a}$ and $a \mathsf{T} \mathsf{R}^+ \subset D_a$, there is $a \mathsf{T} \theta \in \operatorname{Int} V_n$ for some $\theta \in \mathsf{R}^+$; then there exists an open neighbourhood U of a with $U \mathsf{T} \theta \subset \operatorname{Int} V_n$ and $U \times (\theta) \subset$ *domain* T. For all $x \in U$ one has

$$x \mathsf{T} \theta \in \operatorname{Int} V_n \subset V_n ,$$

and therefore from 2.6,

$$L_x^+ = L_{x\mathsf{T}\theta}^+ \subset L_a^+ = T .$$

Since V_n is compact +invariant, L_x^+ is non-void invariant, and thus $L_x^+ = T$. This concludes the proof.

2.8. *Example.* Even in the very special situation of 2.7, the maximal region of attraction need not be open in P. Indeed, for the ld system of (3), trajectories inner to K spiral towards K, and the outer trajectories spiral away from K; the maximal region of attraction is

$$\{z \in \mathsf{R}^2 : 0 < |z| \leqq 1\} .$$

2.9. *Theorem. In a continuous ld system T on a dichotomic carrier P, let L_a^+ be a compact non-critical +limit set of a non-cyclic +trajectory $a \mathsf{T} \mathsf{R}^+$, and denote by D_a that component of $P - L_a^+$ which contains a. If there are no +asymptotic points in D_a near L_a^+, then L_a^+ is ao+ stable relative to* $\overline{D_a}$.

Remark. The heuristics of this theorem are perhaps obvious, at least in $P = \mathsf{R}^2$ as carrier: in applying 2.5, no cycle can intervene between a and its +limit set. The proof proper is rather more complicated, and proceeds in another direction.

Proof of 2.9. 2.9.1. Choose a non-critical $b \in L_a^+$, and perform the construction of 2.1, obtaining D_a, C, G_n, C_n, V_n as indicated there. For definiteness, let

$$a \mathsf{T} \theta_n \in C \quad \text{with} \quad \theta_1 < \theta_n < \theta_{n+1} \to +\infty$$

be all the intersections of $a \mathsf{T} \mathsf{R}^+$ with C. The closed curve C_n then consists of the trajectorial arc $a \mathsf{T} [\theta_n, \theta_{n+1}]$ and of the arc on C between $a \mathsf{T} \theta_n$ and $a \mathsf{T} \theta_{n+1}$ (cf. VII,4.13).

According to 2.2, it may be assumed that V_1 is compact, and also that there are no +asymptotic points in $V_1 - L_a^+$ (otherwise one could abbreviate C). Next, let

$$S = C \cap (\overline{G_2} - G_3)$$

be the arc on C between $a \top \theta_2$ and $a \top \theta_3$. Finally, choose any integer $n > 1$ and define W by

$$W = V_1 - G_{2n} = \overline{G_2} - G_{2n} \tag{5}$$

(since $V_1 = \overline{G_2} \cap \overline{D_a}$, and $P - G_{2n} \subset D_a$ from 1.20). Observe that W is compact, and that from $L_a^+ \subset \overline{G_{2n+2}} \subset G_{2n}$ there follows

$$W = V_1 - G_{2n} \subset V_1 - L_a^+ ,$$

so that by assumption there are no +asymptotic points in W. Also recall that V_1 and G_{2n} are +invariant, so that W is t-convex (cf. 2.2, VII,4.13, II,5.9). Finally, from (5) and VII,4.12 it follows that W is connected. This concludes the preparations.

2.9.2. Let Q be the set of all $x \in S$ such that $x \top \theta \in P - W$ for some $\theta \in \mathsf{R}^+$. Then $a \top \theta_2$ is in Q, since $a \top \theta_{2n+1}$ is in G_{2n} and hence not in $W \subset P - G_{2n}$. From continuity of $\top$ and openness of $P - W$, the set Q is open in S; since $Q \subset S \subset C$ and C is a transversal, Q contains no critical points.

Take any $x \in S - Q$, so that $x \top \mathsf{R}^+ \subset W$. From the properties of W, S, x and from 1.13 it follows that L_x^+ is a cycle in W. Using (5),

$$\operatorname{Fr} W = \operatorname{Fr} \left(\overline{G_2} \cap (P - G_{2n})\right) \subset \left(\operatorname{Fr} \overline{G_2}\right) \cup \operatorname{Fr} \left(P - G_{2n}\right) = C_2 \cup C_{2n} ;$$

hence from VII,4.7, the cycle L_x^+ does not intersect Fr W, and therefore

$$L_x^+ \subset \operatorname{Int} W .$$

For the same reason, $x \notin L_x^+$; then 2.8 yields a neighbourhood U of x in S such that $L_y^+ = L_x^+$ for all $y \in U$. For arbitrarily large $\theta \in \mathsf{R}^+$ one then has that $y \top \theta$ is near L_y^+ and thus within Int W. Since W is t-convex, this implies $y \top \mathsf{R}^+ \subset W$. Thus all $y \in U$ are also in $S - Q$, proving that $S - Q$ is open in S.

Summarizing, both Q and $S - Q$ are open in S. Since Q is non-void and S connected, this yields that $Q = S$. Eliminating the auxiliary concepts, one has the following assertion: To each $x \in S$ there is a $\theta \in \mathsf{R}^+$ with $x \top \theta \notin W$.

2.9.3. Obviously there exists a $\theta_0 \in \mathsf{R}^+$ such that

$$S \top [\theta_0, +\infty) \subset P - W ;$$

in other terms, the property described in 2.9.2 is "uniform" with respect to $x \in S$. This follows directly from compactness of S.

2.9.4. Now set

$$S_0 = S - (a \top \theta_3) = S - \overline{G_3} . \tag{6}$$

From the canonic image of neighbourhoods of S (cf. VII,2.5) it follows easily that $S_0 \top [0, \varepsilon)$ is open in $\overline{G_1}$ for small $\varepsilon > 0$; hence

$$S_0 \top \mathsf{R}^+ = (S_0 \top [0, \varepsilon)) \top \mathsf{R}^+ = \bigcup_{\theta \in \mathsf{R}^+} (S_0 \top [0, \varepsilon)) \top \theta$$

is also open in $\overline{G_1}$, and therefore $W \cap (S_0 \top \mathsf{R}^+)$ is open in W.

Secondly, from 2.9.3,

$$W \cap (S_0 \top \mathsf{R}^+) = W \cap (S_0 \top [0, \theta_0]), \tag{7}$$

and we proceed to prove that this set is closed. Let

$$x_i \top \theta_i \to y$$

with $x_i \in S_0$, $\theta_i \in [0, \theta_0]$, $x_i \top \theta_i \in W$. It may even be assumed that $x_i \to x \in S = \overline{S_0}$ and $\theta_i \to \theta \in [0, \theta_0]$. Now there are two cases: either $x \in S_0$, and then

$$y \leftarrow x_i \top \theta_i \to x \top \theta$$

and $y = x \top \theta$ belongs to $W \cap (S_0 \top [0, \theta_0])$; or $x \in S - S_0$, whereupon from (6) $x = a \top \theta_3$ and $y = x \top \theta = a \top (\theta_3 + \theta)$ is again in $W \cap (S_0 \top \mathsf{R}^+)$ (cf. (7)).

Thus it has been shown that $W \cap (S_0 \top \mathsf{R}^+)$ is both open and closed in a connected set W; since $W \cap (S_0 \top \mathsf{R}^+) \supset S_0$ is non-void, there follows

$$W = W \cap (S_0 \top \mathsf{R}^+),$$

in other words,

$$W \subset S_0 \top \mathsf{R}^+ . \tag{8}$$

2.9.5. The next step shows that no cycles intersect W. Indeed, if $x \in W$ is on a cycle T, then from (8), $x = y \top \theta$ with $y \in S_0$; this would imply $y \in T \cap S_0 \subset T \cap C_2$, contradicting VII,4.7. Using (5) to reformulate, no cycles intersect $V_1 - G_{2n}$ for any $n > 1$; thus no cycle intersects

$$\bigcap (V_1 - G_{2n}) = V_1 \cap \bigcup (P - G_{2n}) = V_1 \cap D_a$$

having applied 1.20, and

$$= \overline{G_1} \cap D_a = \overline{G_1} \cap (\overline{D}_a - L_a^+) = V_1 - L_a^+$$

on using Fr $D_a = L_a^+$ from 1.18.

2.9.6. From the outset it has been assumed that V_1 is compact and that no +asymptotic points are present in $V_1 - L_a^+$. With the result of 2.9.5, this verifies the conditions of 2.5; thus V_1 is a region of attraction for L_a^+. As in 2.3,

$$L_a^+ \subset \overline{G_4} \subset G_2 ,$$

so that $V_1 = \bar{G}_2 \cap \bar{D}_a$ is a neighbourhood of L_a^+ relative to $\bar{D}_a$. This concludes the proof of 2.9.

2.10. Corollary. In a dichotomic carrier P, let L_a^+ be compact non-critical with $a \notin L_a^+$, and let D_a be the component of $P - L_a^+$ with $a \in D_a$. If there are no +asymptotic points in D_a near L_a^+, then there are no critical points in D_a near L_a^+.

3. STABILITY OF CYCLES

Since there are no critical points on or near a cycle, the stability properties of *limit* cycles follow directly from 2.9 : each +limit cycle is *ao*+ stable relative to the appropriate residual domain. To obtain results on general cycles, a topological lemma will be needed; its purpose is to replace the lemma 1.20 applying to +limit sets.

3.1. On a dichotomic set P, let there be given disjoint curves C_n for $1 \leqq n \leqq \infty$ such that

$$\lim C_n = C_\infty ; \tag{1}$$

furthermore, assume that in the decomposition into components corresponding to each C_n,

$$P - C_n = G_n \cup (P - \overline{G_n}) , \tag{2}$$

"monotonicity" obtains, in the following sense:

$$G_1 \supset \ldots \supset G_n \supset G_{n+1} \supset \ldots \supset G_\infty . \tag{3}$$

(According to VII,4.12, this amounts to a particular choice of component of each $P - C_n$.)

3.2. Lemma. Under the assumptions of 3.1,

$$\overline{G_\infty} = \bigcap G_n = \bigcap \overline{G_n} , \tag{4}$$

and $\lim (\overline{G_n} - G_\infty) = C_\infty$ *with* $\overline{G_n} - G_\infty$ *compact for sufficiently large n.*

Proof. 3.2.1. Since the $C_n = \operatorname{Fr} G_n$ are disjoint, from (3)

$$\overline{G_m} = G_m \cup C_m \subset G_n \cup (\overline{G_n} - C_n) = G_n \tag{5}$$

whenever $n < m \leqq +\infty$; in particular,

$$\overline{G_\infty} \subset \bigcap G_n = \bigcap \overline{G_n} . \tag{6}$$

3.2.2. Before proving the converse inclusion, notice the set

$$H = P - \bigcap G_n = P - \bigcap \overline{G_n} .$$

From (3), for each sequence of integers $m_n \to +\infty$, also $H = P - \overline{\bigcap G_{m_n}}$; hence (using the property from Kuratowski, 1950, § 44, III, 1 mentioned in 1.20.3),

$$\operatorname{Fr} H = \operatorname{Fr}(P - \overline{\bigcap G_{m_n}}) = \operatorname{Fr} \bigcup (P - \overline{G_{m_n}}) \subset \overline{\bigcup \operatorname{Fr}(P - \overline{G_{m_n}})} = \overline{\bigcup C_{m_n}} .$$

Therefore

$$\operatorname{Fr} H = \bigcap_{\{m_n\}} \overline{\bigcup_n C_{m_n}} \subset C_\infty , \tag{7}$$

with the intersection taken over all sequences of integers $m_n \to +\infty$; the last inclusion follows easily from (1).

3.2.3. From (6), H is a subset (obviously non-void) of the connected open set $P - \overline{G_\infty}$, and from (7)

$$\operatorname{Fr} H \subset C_\infty = \operatorname{Fr}(P - \overline{G}_\infty) .$$

Hence $H = P - \overline{G_\infty}$, i.e. (4).

Finally,

$$\bigcap (\overline{G_n} - G_\infty) = \bigcap \overline{G_n} - G_\infty = \overline{G_\infty} - G_\infty = C_\infty .$$

Since P is locally compact and C_∞ compact non-void, the last assertion of 3.2 follows immediately.

3.3. Cycle stability theorem. Let T be a cycle of a continuous ld system T *on a dichotomic carrier P, and let D be either of the two components of $P - T$. Then either*

1. Relative to $\overline{D}$, T is an ao+ stable or ao− stable limit cycle, and there are no cycles in D near T; or

2. T is orbitally stable relative to $\overline{D}$, and there exist cycles arbitrarily near T in D.

Proof. 3.3.1. First observe that if T' is any cycle in D, and D' the component of $P - T'$ containing T, then $D' \cap \overline{D}$ is an invariant neighbourhood of T in $\overline{D}$. Hence from 3.2 (and VII,4.12 and VII,4.15), if there are cycles sufficiently close to T in D, then T has arbitrarily small invariant neighbourhoods in $\overline{D}$, i.e. T is orbitally stable relative to $\overline{D}$ (cf. IV,4.9).

3.3.2. Next, let there be no cycles in D near T; then some neighbourhood V of T has no critical points nor cycles in $V \cap D$. Take any $a \in T$ and transversal C to a as in VII,2.6. If λ denotes the primitive period of T, then

$$x \mathsf{T} \lambda \to a \mathsf{T} \lambda = a \quad \text{as} \quad x \to a ;$$

from the canonic neighbourhood of a it also follows that to each $x \in C \cap D$ sufficiently near a there is a $\theta \in \mathsf{R}^+$ with $x \mathsf{T} \theta \in C \cap D$ and small $|\theta - \lambda|$. There results a closed curve C_x, consisting of $x \mathsf{T} [0, \theta]$ and of the subarc of C between x and $x \mathsf{T} \theta$.

On taking $x \in C \cap D$ sufficiently near a, it may even be assumed that the component C_x of $P - C_x$ which contains T has $\overline{G_x} \cap \bar{D}$ a compact subset of V. This follows from 3.2 with $G_\infty = P - \bar{D}$, performing an appropriate sequence of the preceding constructions.

3.3.3. According to VII,4.7, G_x is +invariant or −invariant (it may even be invariant, iff $x = x \mathrm{T} \theta$); for definiteness assume it is the former. Then $\overline{G_x} \cap \bar{D}$ is compact +invariant without critical points (as a subset of V), and from the Poincaré theorem 1.13, L_x^+ is a cycle. Since T is the only cycle in $V \cap \bar{D}$, it follows that $L_x^+ = T$; recalling that $x \in D \subset P - T$, merely apply 2.9 to obtain 1 and hence conclude the proof of 3.3.

3.4. Corollary. Each isolated cycle on a dichotomic carrier is either ao+ stable or ao− stable or, finally, ao+ stable relative to one and ao− stable relative to its second residual domain. A cycle is of the latter type iff it is neither o+ stable nor o− stable.

3.5. Cycles of the last-described type are usually called *semi-stable*. Next, assume given two distinct cycles T and T', whose residual domains G, G' respectively have

$$T \subset G', \quad T' \subset G.$$

From VII,4.12, $G \cap G'$ is a component of both $G - T'$ and $G' - T$. If $G \cap G'$ has compact closure and contains no critical points nor cycles, then T and T' are called *consecutive cycles* (and $G \cap G'$ the region *between* them).

3.6. Poincaré consecutivity theorem. In a continuous ld system on a dichotomic carrier, the region G between consecutive cycles is a region of attraction for precisely one of the cycles; in particular, the cycles are not both ao+ stable nor both ao− stable relative to G.

Proof. Let G be the region between consecutive cycles T_1 and T_2; for $k = 1, 2$ let G_k be the set of $x \in G$ with $L_x^+ = T_k$, so that the G_k are disjoint. From 1.13 and consecutivity, $G = G_1 \cup G_2$; according to 2.7, both G_k are open. Since G is a component, it is connected, and it follows that G coincides with precisely one G_k and the other is void. This concludes the proof.

It may be observed that in the definition of consecutive cycles, it would have sufficed to assume that the region between them contains no cycles nor +asymptotic points (cf. 2.9−10).

4. ISOLATED CRITICAL POINTS

This section considers the behaviour of a continuous ld system T on a 2-manifold P, in the neighbourhood of an isolated critical point. Since only local properties are considered, it is not necessary to assume that the carrier is dichotomic; howerer many of the proofs are simplified by assuming $P = \mathsf{R}^2$ (without any loss of generality, cf. VII, 4.3).

Proposition 4.1 establishes a classification of $o+$ stable isolated critical points, and suggests an analogy between these and cycles as described in 3.3. The remaining portion of this section is devoted to proving and exploiting proposition 4.3. It may be worth nothing that an $o+$ stable point is necessarily critical (cf. IV,4.12 and II,5.6), and that each $ao+$ stable point is $+$asymptotic isolated critical.

4.1. Proposition. Let a be an $o+$ stable isolated critical point (of a continuous ld system on a 2-manifold). Then a satisfies precisely one the following conditions.

1. a is $+$asymptotic, and is then called a (topological) focus.

2. Some neighbourhood of a consists of cycles and a; then a is called a centre.

3. There are both cycles and non-cyclic trajectories arbitrarily near a; then a is called a centre-focus.

Furthermore, a is $ao+$ stable iff 1 holds, and orbitally stable iff 2 or 3 holds.

Proof. Assume that, for a point a as indicated, neither 2 nor 3 occur; then there exists a compact $+$invariant neighbourhood U of a which intersects no cycles. Take any open $+$invariant neighbourhood $V \subset U$ of a (cf. IV,2.13) and any $x \in U$. If one had $x \mathsf{T} \mathsf{R}^+ \subset U - V$, then from the Poincaré theorem 1.13, L_x^+ would be a cycle in U : a contradiction. Thus $x \mathsf{T} \theta \in V$ for some $\theta \in \mathsf{R}^+$, and therefore $L_x^+ \subset \overline{V}$ from $+$invariance; and since $\overline{V}$ may be taken arbitrarily small, it follows that $L_x^+ = (a)$ whenever $x \in U$. This proves both 1 and the assertion on $ao+$ stability.

In case 2 and 3 observe that, on taking a $+$invariant neighbourhood $U \approx \mathsf{R}^2$ of a, any cycle intersecting U must lie within U and contain a in its interior (cf. VII,4.8); hence, the assertion on orbital stability. This concludes the proof of 4.1.

4.2. Example. The local behaviour of the dynamical system as described in 4.1 is easily demonstrated. Take any closed set F in $[0, 1] \subset \mathsf{R}^1$, and choose any map $f:[0, 1] \to \mathsf{R}^-$ of class C^1 having F as its zero-point set. For example, in any open-interval (α, β) contiguous to F set

$$f(x) = -(\beta - \alpha) \exp - \left(\frac{1}{(x-\alpha)^2} + \frac{1}{(x-\beta)^2}\right).$$

Then define, in the interior of the unit circle in R^2, a differential ld system T by means of the equation

$$\frac{\mathrm{d}z}{\mathrm{d}\theta} = z(i + f(|z|))\,;$$

this corresponds to the "polar" equation $\mathrm{d}\varrho/\mathrm{d}\varphi = \varrho f(\varrho)$. The origin is an isolated critical point, and since $\mathrm{d}\varrho/\mathrm{d}\varphi \leq 0$, the origin is $o+$ stable. The cycles of T are precisely those circles $|z| = \varrho$ which have $f(\varrho) = 0$. The cases 4.1.1–3 may then be illustrated by taking as the set F, in turn, the empty set, the complete segment $[0, 1]$ and any set with 0 as accumulation point.

4.3. Proposition. Each o+ stable isolated critical point, not a centre, possesses arbitrarily small neighbourhoods bounded by a closed transversal curve.

Proof. 4.3.1. Let a be a critical point as indicated. In particular, there exists a neighbourhood $G \approx \mathsf{R}^2$ of a which contains no critical points other than a; and obviously G may be taken as small as one pleases. From the assumptions on a and 4.1, there are two cases; first take that of a being a focus.

4.3.2. Take a +invariant neighbourhood $U \subset G$ of a which is also a compact region of attraction for a (cf. 4.1.1); and then a +invariant open neighbourhood V of a with $\overline{V} \subset \operatorname{Int} U$. Also set $Q = U - V$, a compact subset of U; the first step will consist of proving that

$$Q \,\dot{\mathsf{T}}\, \mathsf{R}^+ = U - (a)\,. \tag{1}$$

4.3.3. Since U is +invariant and (a) invariant, $U - (a)$ is +invariant; and therefore

$$Q \,\mathsf{T}\, \mathsf{R}^+ = (U - V) \,\mathsf{T}\, \mathsf{R}^+ \subset (U - (a)) \,\mathsf{T}\, \mathsf{R}^+ = U - (a)\,,$$

one inclusion needed for (1).

For the converse inclusion, take any $x \in U - (a)$. If $x \notin V$ then, of course,

$$x \in U - V \subset (U - V) \,\mathsf{T}\, \mathsf{R}^+ = Q \,\mathsf{T}\, \mathsf{R}^+\,.$$

Consider, therefore, the case that $x \in V - (a)$. Principally, there are two possibilities.

The first of these is that

$$x \,\mathsf{T}\, \mathsf{R}^- \subset V\,. \tag{2}$$

Since $V \subset U$ and U is compact and without cycles, from 1.13 it follows that $a \in L_x^-$; thus there exist $\theta_n \to -\infty$ in R^- with $x \,\mathsf{T}\, \theta_n \to a$. Since $x \neq a$, there is a +invariant neighbourhood W of a with $x \notin W$. Then $x \,\mathsf{T}\, \theta_n \in W$ for large n; but

$$x = (x \,\mathsf{T}\, \theta_n) \,\mathsf{T}\, (-\theta_n) \in W \,\mathsf{T}\, (-\theta_n) \subset W \,\mathsf{T}\, \mathsf{R}^+ = W\,,$$

contradicting the choice of W. One concludes that (3) is impossible, so that the only possible case for x is that $x \,\mathsf{T}\, \theta \notin V$ for some $\theta \in \mathsf{R}^-$. Taking $|\theta|$ the least possible, there results

$$x \,\mathsf{T}\, \theta \in \operatorname{Fr} V, \quad \theta \in \mathsf{R}^-\,.$$

Since $\overline{V} \subset U$ with V open, one has $\operatorname{Fr} V \subset U - V = Q$, and therefore

$$x = (x \,\mathsf{T}\, \theta) \,\mathsf{T}\, (-\theta) \subset Q \,\mathsf{T}\, (-\theta) \subset Q \,\mathsf{T}\, \mathsf{R}^+\,.$$

Summarizing, $x \in U - (a)$ implies $x \in Q \,\mathsf{T}\, \mathsf{R}^+$; this gives the second inclusion required in proving (1).

4.3.4. Since U is a region of uniform attraction (cf. VI,3.12), there exists a $\lambda > 0$ such that $U \text{T} [\lambda, +\infty) \subset V$; this implies

$$Q \text{T} [\lambda, 3\lambda] \subset U \text{T} [\lambda, +\infty) \subset V \subset P - Q ,$$

so that Q and $Q \text{T} [\lambda, 3\lambda]$ are disjoint. Also note that $\alpha_x = +\infty$ for all $x \in U$ since U is compact +invariant (IV,1.10). Thus Q and λ satisfy the conditions of VI.2.21; in particular, there exists a continuous map $\varphi : Q \text{T} [0, 2\lambda] \to \mathsf{R}^+$ with the following properties:

$$\varphi \mid Q = 0 , \quad \varphi \mid (Q \text{T} [\lambda, 2\lambda]) = 1 ,$$

and the set

$$S = \{x \in Q \text{T} [0, 2\lambda] : \varphi(x) = \tfrac{1}{2}\}$$

is a compact section.

4.3.5. Now extend φ to a map $\psi : P \to \mathsf{R}^+$ by the following assignments:

$$\psi \mid (P - U) = 0 ,$$
$$\psi \mid (Q \text{T} (2\lambda, +\infty)) = \psi(a) = 1 .$$

Since $Q \text{T} (2\lambda, +\infty) \cup (a)$ is a neighbourhood of a (cf. (1)), ψ is continuous at a, and hence also in

$$(Q \text{T} [0, 2\lambda]) \cup (Q \text{T} (2\lambda, +\infty)) \cup (a) = (Q \text{T} \mathsf{R}^+) \cup (a) = U .$$

Secondly, since

$$Q \supset U - V \supset \text{Int}\, U - \overline{V}$$

and $\varphi \mid Q = 0$, ψ is continuous at $\text{Fr}\, U$, and hence also in $(P - U) \cup U = P$. Finally, from the extension there follows

$$\{x \in P : \psi(x) = \tfrac{1}{2}\} = \{x \in Q \text{T} [0, 2\lambda] : \varphi(x) = \tfrac{1}{2}\} = S .$$

4.3.6. Since S is a compact section, it has a finite system of components, each of which is then a compact connected section. According to VII,1.6, the non-singleton components are transversal curves.

Obviously S separates G into

$$G \cap \{x : \psi(x) < \tfrac{1}{2}\} \quad \text{and} \quad G \cap \{x : \psi(x) > \tfrac{1}{2}\} ;$$

the former set contains Q (non-void, from (1)), the latter contains a. Since $G \approx \mathsf{R}^2$ and R^2 is not separated by any finite system of disjoint arcs and singletons, there exists a closed transversal curve $C \subset S$. From the original assumption on G and VII,4.8, the inner region of C (interpreted in $G \approx \mathsf{R}^2$) must contain a. This completes the proof in the case of a a focus.

4.3.7. Finally, let a be a centre-focus. Take a noncyclic non-critical trajectory $x \text{T} \mathsf{R}^1 \subset G$; according to 1.13 and 1.7, L_x^+ and L_x^- are distinct cycles; assume that for example, L_x^+ is in the inner region of L_x^- (again, interpreted in $G \approx \mathsf{R}^2$). Now collapse the inner region of L_x^+ into a point x', preserving $P - G$ intact; there results a new continuous ld system on a 2-manifold again, and with x' manifestly a +asymptotic stable isolated critical point. On applying 4.3.2-6 to x' and returning to the original system and carrier, one obtains a closed transversal curve C in the region between the consecutive cycles L_x^+ and L_x^-. As in 4.3.6, C bounds a neighbourhood of a in G. This concludes the proof of 4.3.

4.4. Corollary. An isolated critical point $a \in P$ has arbitrarily small neighbourhoods bounded by transversals iff a is $o+$ or $o-$ stable, and not a centre.

4.5. In 4.3 – 4 there may be recognized a second classification of $o+$ stable isolated critical points. This is brought into relation with that of 4.1 in Table I.

Table I

Type	Orbital stability	Trajectories	Transversals
focus	$ao+$, non $o-$ / $ao-$, non $o+$	no	yes
centre	$o+$, $o-$, non $ao\pm$	yes	no
centre-focus	$o+$, $o-$, non $ao\pm$	yes	yes

The entries in the latter two columns concern existence of arbitrarily small neighbourhoods bounded by closed curves of the indicated types. The no's are consequences of the first non-intersection theorem VII,4.4.

4.6. Theorem. Let T be a continuous ld system on a 2-manifold P. Then near each $ao+$ stable point $a \in P$, T is isomorphic in the category $\mathscr{D}_l^T$ to the differential gd system defined in R^2 by the linear equation

$$\frac{\mathrm{d}z}{\mathrm{d}\theta} = -z\,. \tag{3}$$

Proof. 4.6.1. From 4.3, there exists a compact +invariant neighbourhood U of a bounded by a closed transversal curve C. Let E^2 be the unit disc in R^2, S^1 its boundary, and parametrize $p : \mathsf{S}^1 \approx C$.

The gd system T' associated with (3) may be described by

$$x\,\mathsf{T}'\,\theta = xe^{-\theta} \quad (x \in \mathsf{R}^2,\ \theta \in \mathsf{R}^1)\,;$$

this suggests defining a map $f : \mathsf{E}^2 \to U$ so as to be a morphism between T' and T:

$$f(\sigma e^{-\theta}) = p(\sigma)\,\mathsf{T}\,\theta \quad \text{for} \quad \sigma \in \mathsf{S}^1,\ \theta \in \mathsf{R}^+\,,$$
$$f(0) = a\,.$$

4.6.2. Since U is compact +invariant, $p(\sigma) \top \theta$ is indeed defined for all the indicated σ, θ. Evidently f is continuous at each $x \neq 0$ in E^2. To prove continuity at 0, let $0 \neq x_n \to 0$; define $(\sigma_n, \theta_n) \in \mathsf{S}^1 \times \mathsf{R}^+$ by $x_n = \sigma_n e^{-\theta_n}$, so that $\theta_n \to +\infty$. According to VI,3.14, U is a region of uniform attraction to a, so that

$$f(x_n) = p(\sigma_n) \top \theta_n \in U \top \theta_n \to a = f(0) .$$

Thus f is continuous throughout E^2.

4.6.3. Next, show that f is $1-1$. Obviously $f(x) \neq f(0)$ whenever $0 \neq x \in \mathsf{E}^2$. If

$$f(\sigma e^{-\theta}) = f(\sigma' e^{-\theta'})$$

with both (σ, θ) and (σ', θ') in $\mathsf{S}^1 \times \mathsf{R}^+$ and, for example, $\theta' \geqq \theta$, then

$$p(\sigma) = p(\sigma') \top (\theta' - \theta) .$$

$p(\sigma) \ni C \in p(\sigma')$ and the non-intersection theorem VII,4.4 then imply $\theta' = \theta$ and $p(\sigma) = p(\sigma')$, i.e. $\sigma = \sigma'$, as required.

4.6.4. Since E^2 is compact, 4.6.2–3 yield that f is a homeomorphism. From the preservation of domain theorem it follows that

$$\mathrm{Fr}\, f(\mathsf{E}^2) = f(\mathrm{Fr}\, \mathsf{E}^2) = f(\mathsf{S}^1) = C = \mathrm{Fr}\, U ,$$

so that $f : \mathsf{E}^2 \approx U$. This concludes the proof of 4.6.

4.7. Corollary. In the situation of 4.6 there exists a (Lyapunov function) continuous map $\varphi : P \to \mathsf{R}^+$ with a unique zero-point at a, and such that

$$\frac{\partial}{\partial \theta} \varphi(x \top \theta) = -\varphi(x \top \theta)$$

for x in a neighbourhood of a and all $\theta \in \mathsf{R}^+$. Furthermore, near a, each level curve is a simple closed curve, and these are nested around a.

Proof. With the notation of the preceding proof, set $\varphi(x) = |f^{-1}(x)|$ for $x \in U$ and $\varphi \mid (P - U) = 1$.

4.8. According to 4.6, in terms of the topologico-dynamical category $\mathscr{D}_i^T$, there is precisely one type of $ao+$ stable point, namely the (topological) focus of (3). From 4.2 it is seen that there are many types of centre-focus (depending on the choice of F, but this does not exhaust the variety). Similarly, there are several types of centers; this is obvious from the two examples in VII,4.17.2, which also suggest that the different types may be made to correspond with the limiting behaviour of the primitive periods as the center is approached.

Finally, consider non-$o+$ stable isolated critical points.

4.9. Proposition. Each non-o+ stable isolated critical point is −asymptotic.

Proof. 4.9.1. Let a be a critical point as indicated, so that there is a neighbourhood U of a, a sequence $x_i \to a$ in P, and $\theta_i \in \mathsf{R}^+$ with $x_i \top \theta_i \notin U$. Take a second neighbourhood V of a with $\mathsf{E}^2 \approx V \subset U$ not containing cycles nor critical points other than a (no cycles as a would then be orbitally stable, see 4.5). Since $x_i \to a$, it may be assumed that $x_i \in \operatorname{Int} V$; then let $\theta_i' \in \mathsf{R}^+$ be the least with $x_i \top \theta_i' \in \operatorname{Fr} V$, so that

$$x_i \top [0, \theta_i'] \subset V. \tag{4}$$

Since $\operatorname{Fr} V \subset V \approx \mathsf{E}^2$ are compact, there is $x_j \top \theta_j' \to x' \in \operatorname{Fr} V$ for some subsequence; from $a \in \operatorname{Int} V$ it then follows, of course, that $a \neq x'$.

4.9.2. The next step is to show that $\theta_j' \to +\infty$. Assume the contrary; then $\theta_k' \to \theta$ for some subsequence and $\theta \in \mathsf{R}^+$. But then, from criticality of $a = \lim x_k$,

$$a = a \top \theta \leftarrow x_k \top \theta_k' \to x',$$

contradicting $a \neq x'$.

4.9.3. Now prove $x' \top \mathsf{R}^- \subset V$. Take any $\theta \in \mathsf{R}^-$ with $x' \top \theta$ defined. From 4.9.2, $-\theta_j' \leqq \theta \leqq 0$ for large j, i.e.

$$0 \leqq \theta_j' + \theta \leqq \theta_j';$$

then, using (4),

$$x' \top \theta \leftarrow (x_j \top \theta_j') \top \theta = x_j \top (\theta_j' + \theta) \in V.$$

Thus indeed $x' \top \mathsf{R}^- \subset V$, and therefore

$$L_{x'}^- \subset \overline{x' \top \mathsf{R}^-} \subset V.$$

From the Poincaré theorem 1.13 it follows that $L_{x'}^-$ contains a critical point or a cycle, so that, by construction of V in 4.9.1, $a \in L_{x'}^-$. Together with $a \neq x'$ this concludes the proof of 4.9.

4.10. Corollary. An isolated critical point neither o+ stable nor o− stable is both +asymptotic and −asymptotic.

4.11. Corollary. If a is isolated critical and

$$a \in L_x^+, \quad (a) \neq L_x^+,$$

then a is both +asymptotic and −asymptotic.

4.12. For isolated critical points, 4.1 and 4.9 may be summarized by the following implications:

$$\text{non } -\text{asymptotic} \Rightarrow o+ \text{ stable} \Rightarrow ao+ \text{ stable or } o \text{ stable}.$$

CHAPTER IX

Flows

For the motivation of the concept of a flow in the present context, see I,6.3 and examples 1.9.1 and 1.9.5 to follow. Section 1 introduces and illustrates flows over the parameter domains defined in 1.1; the illustration are intended as a common generalization of the parameter domains R^1 and R^+ of global dynamical systems, and also of C^1 and C^+. Some special types of flow (admitting a period, stationary and invariant relative to a map) are reviewed in Section 2. This forms a preliminary to Section 3, where the principal results of this chapter are presented – the existence theorems for periodic solutions. Applications to dynamical system theory are then given in Section 4.

Originally, the applications were obtained first, in the form of critical point existence theorems for dynamical systems (for example 4.3 in Hájek 4, 1964). It was then realized that the essential point of difference between dynamical systems and flows, the group property, was not used in a significant portion of the proof. The actual divorce from dynamical systems was carried out in Hájek 10 (1965), on which most of this chapter is based.

For most of the other results it is rather difficult to give specific references. Thus, in differential equation theory, most of Section 2 is probably familiar, and corollary 3.8 (via 3.3 and 2.5.1) is definitely known. In the same context, proposition 3.17 would seem to belong to the theory of point mappings, as the term is used in Neymark (1963).

1. FLOWS AND SEMI-FLOWS

1.1. A parameter domain will mean a quadruple $(R, +, \geqq, t)$ with R a set and $+, \geqq, t$ structures on R having the following six properties.

1. The $+$ is a semi-group operator on R, i.e. a binary associative operator on R; furthermore, the existence of a zero $o \in R$ is required:

$$\alpha + o = o + \alpha = \alpha \quad \text{for all} \quad \alpha \in R\,.$$

2. The $\geqq$ is a quasi-order in R, i.e. a reflexive and transitive relation on R.

3. The t is a topology on R.

4. The triple $(R, +, \geqq)$ is a quasi-ordered semigroup.

5. The triple $(R, +, t)$ is a topological semi-group.
6. The triple $(R, \geqq, t)$ is a topological quasiordered space.

1.2. The three compatibility conditions 4–6 may be described at greater length as follows (lower-case Greek letters will consistently be used to denote elements of R). For 4:

$$\alpha \geqq \beta \quad \text{and} \quad \alpha' \geqq \beta' \quad \text{implies} \quad \alpha + \alpha' \geqq \beta + \beta' .$$

For 5: the operator $+$, interpreted as a mapping $R \times R \to R$, is continuous in the induced and given topologies; in other words,

$$\alpha_i \to \alpha \quad \text{and} \quad \beta_i \to \beta \quad \text{implies} \quad \alpha_i + \beta_i \to \alpha + \beta .$$

For 6: the set $\{(\alpha, \beta) \in R \times R : \alpha \geqq \beta\}$ is closed in $R \times R$; i.e.

$$\alpha_i \to \alpha , \quad \beta_i \to \beta , \quad \alpha_i \geqq \beta_i \quad \text{imply} \quad \alpha \geqq \beta .$$

1.3. To avoid needlessly pedantic constructions, the situation described in 1.1 will often be abbreviated to the clause "R is a parameter domain". If $+$ is a group operator (i.e. if $(R, +)$ is a group), then R itself will be called a group; similarly, R will be said to be partially ordered if $\geqq$ is a partial ordering, and to be discrete if the topology t is discrete. If $\geqq$ is the maximal relation on R,

$$\alpha \geqq \beta \quad \text{for all} \quad \alpha, \beta \in R ,$$

then R will be termed *unordered*. (It should be remarked that, in the category of partially ordered sets, this term is sometimes used to describe the minimal reflexive relation, viz. equality; this meaning will not be employed here.)

In connection with the semi-group operator, additive notation will be used, though commutativity is not assumed. In particular, for $\alpha \in R$ and positive integral n one defines

$$n\alpha = \alpha + \ldots + \alpha \quad (n \text{ terms}) ,$$
$$0\alpha = o .$$

1.4. Examples.

1.4.1. The standard parameter domains are R^1, R^+, C^1, C^+ with their natural additive structure, partial order and topology. Other examples may be obtained from the corresponding direct products; for example in R^2 with the additive structure and topology of the complex numbers, the product partial order $\geqq$ is defined by

$$\alpha \geqq \beta \quad \text{iff} \quad \operatorname{Re} \alpha \geqq \operatorname{Re} \beta \quad \text{and} \quad \operatorname{Im} \alpha \geqq \operatorname{Im} \beta ,$$

with the latter two relations interpreted in R^1.

1.4.2. An abstract quasi-ordered semi-group may be identified, in the obvious manner, with a discrete parameter domain. A less careless formulation would be to say that the category of quasi-ordered semi-groups may be injected into that of parameter domains. In a similar sense, a topological semi-group may be identified with an unordered parameter domain; a useful instance of this is the parameter domain R^1 taken unordered.

1.4.3. Let R be an abstract set with distinguished element $o \in R$, and define a binary operator $+$ in R by

$$\alpha + \beta = \begin{cases} \alpha & \text{if} \quad \alpha \neq o\,, \\ \beta & \text{if} \quad \alpha = o\,. \end{cases}$$

It is then readily verified that $+$ is an associative operation with o as zero element, so that $(R, +)$ is a semigroup in the sense of 1.1.1. The set R taken with this operator $+$, unordered and discrete, constitutes a parameter domain. A special case of this situation occurs on taking a singleton for R.

1.4.4. In a similar vein, into any complete distributive lattice one may introduce a natural ("order") topology and a more or less trivial semi-group operation, the lattice join $\vee$ (Birkhoff, 1948, IX, § 11), obtaining a parameter domain as defined above.

1.5. Lemma. Let R be a partially ordered parameter domain. Then R is a T_2 space; and, for each $\alpha \geqq o$, the cyclic sub-semi-group $\{n\alpha : n \in \mathsf{C}^+\}$ is discrete.

Proof. R is a T_2 space iff the diagonal of $R \times R$, i.e. the set

$$\{(\alpha, \alpha) : \alpha \in R\}\,, \tag{1}$$

is closed (Bourbaki, 1951, I, § 8,6). By assumption, the relation $\geqq$ is anti-symmetric; this may also be formulated by saying that (1) is the intersection of

$$\{(\alpha, \beta) : \alpha \geqq \beta\} \quad \text{and} \quad \{(\alpha, \beta) : \beta \geqq \alpha\}\,. \tag{2}$$

From 1.1.6 (or rather 1.2), the sets (2) are closed. This proves the first assertion.

Next, assume that the indicated semi-group is not discrete. Then there exist k_n, k in C^+ with

$$k_n\alpha \to k\alpha\,, \quad k_n \to +\infty\,.$$

Take any $m \geqq k$ in C^+; then $k_n\alpha \geqq m\alpha$ for large n, and hence

$$k\alpha \leftarrow k_n\alpha \geqq m\alpha \geqq k\alpha\,.$$

Therefore $m\alpha = k\alpha$ for all $m \geqq k$, and $\{n\alpha : n \in \mathsf{C}^+\} = \{n\alpha : 0 \leqq n \leqq k\}$ is finite and hence discrete. This contradiction establishes the second assertion and concludes the proof of 1.5.

1.6. Before formulating the principal definition of this chapter it will be useful to introduce some notational conventions.

The composition of two maps $f : A \to B$ and $g : B \to C$, hereto denoted by gf and defined by

$$g f(x) = g(f(x)) \quad \text{for} \quad x \in A\,,$$

will occasionally be denoted by $g \circ f$; and brackets will often be omitted. The identity map of a (previously given) set A will be denoted by 1,

$$1x = x \quad \text{for} \quad x \in A\,.$$

Now assume given a set P, a parameter domain R, and a map

$$T : P \times \{(\alpha, \beta) \in R \times R : \alpha \geqq \beta\} \to P\,.$$

With each pair $\alpha \geqq \beta$ in R one may then associate a map $P \to P$ to be denoted, standardly by ${}_\alpha T_\beta$ and defined thus:

$${}_\alpha T_\beta x = T(x, \alpha, \beta) \quad \text{for} \quad x \in P\,.$$

Essentially, ${}_\alpha T_\beta$ is the partialization of T to the set $(\textit{domain}\ T) \cap (P \times (\alpha, \beta))$; in particular, if P is topological and T continuous, then each ${}_\alpha T_\beta$ is also continuous.

1.7. Definition. A mapping T is termed a *semi-flow on P over R* if P is topological space, R is a parameter domain, and T has the following three properties:

1. $T : P \times \{(\alpha, \beta) \in R \times R : \alpha \geqq \beta\} \to P$ is a continuous map, in the given and induced topologies.

2. T has the initial value property

$${}_\alpha T_\alpha = 1 \quad \text{for} \quad \alpha \in R\,.$$

3. T has the composition property

$${}_\alpha T_\beta \circ {}_\beta T_\gamma = {}_\alpha T_\gamma \quad \text{whenever} \quad \alpha \geqq \beta \geqq \gamma \text{ in } R\,.$$

If R is unordered, then T is called a *flow*; if R is discrete, then T is also termed *discrete*.

For fixed $x \in P$, a semi-flow T on P defines a continuous map $T_o x : \{\alpha \in R : : \alpha \geqq o\} \to P$ by

$$(T_o x)(\theta) = {}_\theta T_o x \quad \text{for} \quad \theta \geqq o \text{ in } R\,;$$

it is called the *solution* of T through x. Again $T_o x$ may be interpreted as the partialization of T to $(\textit{domain}\ T) \cap ((x) \times \{\alpha : \alpha \geqq o\} \times (o))$. It should be emphasized that, according to this definition, the map $\{\theta : \theta \geqq \alpha\} \to P$ defined by ${}_\theta T_\alpha x$ in general is not a solution of T. It is a solution if $o \geqq \alpha$ since from 1.7.3 there then follows

$${}_\theta T_\alpha x = {}_\theta T_o({}_o T_\alpha x)\,;$$

in particular this obtains if T is a flow. (Also see the "strong" variants of the properties treated in Section 2.)

1.8. Lemma. If T is a flow on P over R, then each ${}_\alpha T_\beta$ is a homeomorphism $P \approx P$ and ${}_\alpha T_\beta^{-1} = {}_\beta T_\alpha$.

Proof. Since R is unordered, ${}_\alpha T_\beta$ is defined for all pairs α, β in R; and then, from 1.7.2–3,

$$ {}_\alpha T_\beta \circ {}_\beta T_\alpha = {}_\alpha T_\alpha = 1 \,, \quad {}_\beta T_\alpha \circ {}_\alpha T_\beta = {}_\beta T_\beta = 1 \,. $$

1.9. Examples.

1.9.1. Let G be an open set in R^n, $f : G \times \mathsf{R}^n \to \mathsf{R}^n$ continuous, and assume that the differential equation

$$ \frac{\mathrm{d}x}{\mathrm{d}\theta} = f(x, \theta) \tag{3} $$

has unicity and extendability of solutions (cf. I,1.6, I,1.11). Take any $(x, \alpha, \beta) \in \in G \times \mathsf{R}^1 \times \mathsf{R}^1$, determine the unique characteristic solution $y(\cdot)$ of (3) with initial condition $y(\beta) = x$, and set

$$ T(x, \alpha, \beta) = y(\alpha) \,. $$

Then T is a flow on G over unordered R^1; this follows from I,6.3 (or, in detail, from I,1.9 and I,1.13). Flows of this type will be called *differential.* Observe that the concept of a characteristic solution of (3) coincides with that of a solution of T; this will be used later on several occasions.

1.9.2. A related example is the following. In a Banach space P let

$$ \frac{\mathrm{d}x}{\mathrm{d}\theta} = A(\theta)\, x \,, \quad (x, \theta) \in P \times \mathsf{R}^1 \,, $$

be a (homogeneous linear) differential equation with $A(\theta)$ a bounded linear operator $P \to P$ depending continuously on $\theta \in \mathsf{R}^1$. For α, β in R^1 let $V(\alpha, \beta) = U(\alpha)\, U^{-1}(\beta)$ be the corresponding resolving operator (Kreĭn, 1964, III, § 1.5; $U(\alpha)$ is bounded linear). Then, essentially, V is a flow T on P over unordered R^1:

$$ {}_\alpha T_\beta x = V(\alpha, \beta)\, x \,. $$

Analogous results may be obtained in significant cases even if the linear operator $A(\theta)$ is unbounded (*loc. cit.*, § 3).

1.9.3. Let $f : \mathsf{R}^2 \times \mathsf{R}^2 \to \mathsf{R}^2$ be a holomorphic function of two variables, and assume that all solutions of

$$ \frac{\mathrm{d}w}{\mathrm{d}z} = f(w, z) \tag{4} $$

are entire functions. Then a definition similar to that of 1.9.1 associates with (4) a flow on R^2 over (unordered) R^2.

1.9.4. Let B be a Boolean σ-algebra, and L_B the set of σ-additive measure functions m on B with $m(0) = 0$. On L_B one then has the obvious vector space structure and partial order, and also a reasonable metric topology obtained by taking the total variation as norm:

$$\|m\| = \sup_{x,y \in B} |m(x) - m(y)| .$$

Furthermore, these three structures are compatible, so that L_B turns out to be a Banach lattice (Birkhoff, 1948, XV, §§ 13–14).

Now, let P be the subspace of L_B consisting of the probability distributions on B, i.e. of the $p \in L_B$ with $0 \leqq p$ and $p(I) = 1$; and take any parameter domain R. Then every semi-flow T on L_B over R, which also has the property that each ${}_\alpha T_\beta$ is linear and maps P into itself, is called a transition operator on L_B; and every solution $T_o p$ with $p \in P$ is called a stochastic process of Markovian type (*loc. cit.*, XVI, § 3) (usually the latter term is reserved for the random variables rather than their distribution functions).

1.9.5. Let T be an abstract gsd system on P; then

$$_\alpha T_\beta = x\,\mathsf{T}\,(\alpha - \beta) \tag{5}$$

defines, for $x \in P$, $\alpha \geqq \beta$ in R^1, a semi-flow on discrete P over discrete R^1. The solutions $T_o x$ of T coincide with the +solutions t_x^+ of T,

$$_\theta T_0 x = x\,\mathsf{T}\,\theta = t_x^+(\theta) \quad \text{for} \quad \theta \geqq 0 .$$

If T is an abstract gd system on P, then (5) defines, on omitting the restriction $\alpha \geqq \beta$, a flow on discrete P over discrete unordered R^1; the solutions of T then coincide with the solutions of T. Similar conclusions hold for continuous global (semi-) dynamical systems on a topological space P.

1.9.6. Take any topological space P; then there is a unique semi-flow on P over the singleton parameter domain (cf. 1.4.3). In the obvious manner, this flow may be identified with the identity map of P, and hence with P itself.

1.9.7. Let $\{f_n \mid n \in \mathsf{C}^+\}$ be a sequence of continuous maps $P \to P$ of a topological space; then

$$_m T_n = f_{m-1} \circ f_{m-2} \circ \dots \circ f_{n+1} \circ f_n, \quad (m > n)$$
$$_n T_n = 1$$

define a discrete semi-flow T on P over C^+. Conversely, every semi-flow T on P over C^+ is of this type: merely take $f_n = {}_{n+1}T_n$. Similarly for C^1 in place of C^+. In either case the semi-flow obtained can be extended to a (uniquely determined) flow iff each f_n is a homeomorphism $P \approx P$.

1.10. The preceding examples suggest – and recourse to 1.7 confirms – that a semi-flow T on P over R is completely determined by the system of maps $P \to P$

$$\{{}_\alpha T_\beta \mid \alpha \geqq \beta \text{ in } R\} . \tag{6}$$

For flows, the considerably smaller system $\{{}_\alpha T_o \mid \alpha \geqq o \text{ in } R\}$ also determines T, since from 1.7.3 and 1.8 there follows

$${}_\alpha T_\beta = {}_\alpha T_o \circ {}_o T_\beta = {}_\alpha T_o \circ {}_\beta T_0^{-1} .$$

Conversely, assume given a topological space P, parameter domain R, and a system (6) of continuous maps $P \to P$; if the system has the initial value property and is compositive as in 1.7.2–3, then it defines a discrete semi-flow T on P over R taken discrete. Whether or not T is a semi-flow on P over R with its given topology is then decided by the manner of dependence of ${}_\alpha T_\beta$ on α and β.

2. SPECIAL TYPES OF SEMI-FLOWS

This section contains the definitions and elementary properties of some special types of semi-flows (cf. 2.1, 2.6, 2.11). In each case the condition imposed on the semi-flow is external, in the sense that for differential flows, associated with a differential equation $dx/d\theta = f(x, \theta)$ as in 1.9.1, the condition can be verified directly on the function f, without any further information concerning the solutions themselves.

2.1. Definition. A semi-flow T on P over R is said to *admit the period* τ if $\tau \geqq o$ in R and

$${}_\alpha T_o \circ {}_\tau T_o = {}_{\alpha+\tau} T_o \quad \text{for} \quad \alpha \geqq o \text{ in } R ; \tag{1}$$

and to *admit the period τ strongly* if

$${}_\alpha T_\beta = {}_{\alpha+\tau} T_{\beta+\tau} \quad \text{for} \quad \alpha \geqq \beta \text{ in } R . \tag{2}$$

2.2. Lemma. The condition (2) *implies* (1); *for flows the two conditions are equivalent.*

Proof. Take any $\alpha \geqq o$ in R, so that $\alpha + \tau \geqq \tau \geqq o$; from (2) and compositivity there then follows

$${}_{\alpha+\tau} T_o = {}_{\alpha+\tau} T_\tau \circ {}_\tau T_o = {}_\alpha T_o \circ {}_\tau T_o ,$$

i.e. (1). Next, let T be a flow with (1); then 1.8 applied to (1) yields

$${}_o T_{\beta+\tau} = {}_o T_\tau \circ {}_o T_\beta ;$$